AF560301

# Fish as Food

# Fish as Food

Naveen Gupta

**Fish as Food**

---

ISBN 978-93-5111-422-2
© Reserved

All Rights Reserved. No Part of this book may be reproduced in any manner without written permission.

Published in 2014 in India by

**RANDOM PUBLICATIONS**

4376-A/4B, Gali Murari Lal, Ansari Road
New Delhi-110 002
Phone : +91-11-43580356, +91-11-23289044
e-mail: randomexports@gmail.com, sales@randompublications.com, info@randompublications.com

Reprint 2022

*Type Setting by:* Friends Media, Delhi-110089
*Printed at :* Replika Press Pvt. Ltd.

# Preface

Fish is a food of excellent nutritional value, providing high quality protein and a wide variety of vitamins and minerals, including vitamins A and D, phosphorus, magnesium, selenium, and iodine in marine fish. Its protein - like that of meat - is easily digestible and favourably complements dietary protein provided by cereals and legumes that are typically consumed in many developing countries. Experts agree that, even in small quantities, fish can have a significant positive impact in improving the quality of dietary protein by complementing the essential amino acids that are often present in low quantities in vegetable-based diets. But recent research shows that fish is much more than just an alternative source of animal protein. Fish oils in fatty fish are the richest source of a type of fat that is vital to normal brain development in unborn babies and infants. Without adequate amounts of these fatty acids, normal brain development does not take place. Closely spaced pregnancies, often seen in developing countries, can lead to the depletion of the mother's supply of essential fatty acids, leaving younger siblings deprived of this vital nutrient at a crucial stage in their growth. This makes fatty fish such as tuna, mackerel and sardine - all of which are commonly available in developing countries - a particularly good choice for the diet of pregnant and lactating women. Research over the past few decades has shown that the nutrients and minerals in fish, and particularly the omega 3 fatty acids found in pelagic fishes, are heart-friendly and can make improvements in brain development and reproduction. This has highlighted the role for fish in the functionality of the human body.

Fish is the most common food to obstruct the airway and cause choking. Choking on fish was responsible for about reported 4,500 accidents in the UK in 1998. A seafood allergy is a hypersensitivity to an allergen which can be present in fish, particularly in shellfish. This can result in an overreaction of the immune system and lead to severe physical symptoms. Most people who have a food allergy also have a seafood allergy. Allergic reactions can result from ingesting seafood, or by breathing in vapours from preparing or cooking seafood. The most severe seafood allergy reaction is anaphylaxis, an emergency requiring immediate attention. It is treated with epinephrine. Some species of fish, notably the puffer fugu used for sushi, and some kinds of shellfish, can result in serious poisoning if not prepared properly. These fish always contain these poisons as a defence against predators; it is not present due to environmental circumstances. Particularly, fugu has a lethal dose of tetrodotoxin in its internal organs and must be prepared by a licensed fugu chef who has passed the national examination in Japan. Ciguatera poisoning can occur from eating larger fish from warm tropical waters, such as sea bass, grouper, and red snapper. Scombroid poisoning can result from eating large oily fish which have sat around for too long before being refrigerated or frozen. This includes scombroids such as tuna and mackerel, but can also include non-scombroids such as mahi-mahi and amberjack. The poison is odourless and tasteless.

A lot of fish eat algae and other organisms that contain biotoxins (defensive substances against predators). Biotoxins accumulated in fish/shellfish include brevetoxins, okadaic acid, saxitoxins, ciguatoxin and domoic acid. Except for ciguatoxine, high levels of these toxins are only found in shellfish. Both domoic acid and ciguatoxine can be deadly to humans; the others will only cause diarrhea, dizzyness and a (temporary) feeling of claustrophobia. Shellfish are filter feeders and, therefore, accumulate toxins produced by microscopic algae, such as dinoflagellates and diatoms, and cyanobacteria. There are four syndromes called shellfish poisoning which can result in humans, sea mammals, and birds from the ingestion of toxic shellfish. These are primarily associated with bivalve molluscs, such as mussels, clams, oysters andscallops. Fish, like anchovies can also concentrate toxins such as domoic acid. If suspected, medical attention should be sought.

The book will be an indispensable source for all professionals, researchers and students in this subject and for anyone working in the related areas for acquiring an up-to-date overviews.

I thank all members of my team who have helped in the preparation of the book. My special thanks go to "Random Publication" who have published the book.

*—Naveen Gupta*

# Contents

# 1

# Introduction

Fish is a food consumed by many species, including humans. The word "fish" refers to both the animal and to the food prepared from it. In culinary and fishery contexts, the term fish also includes shellfish, such as molluscs, crustaceans and echinoderms. Fish has been an important source of protein for humans throughout recorded history. English does not distinguish between fish as an animal and the food prepared from it, as it does with *pig versus pork* or *cow versus beef.* Some other languages do, as in the Spanish *peces versus pescado*. The German *speisefisch* means fish as a food or fish suitable for eating. English also has the term seafood, which covers fish as well as other marine life used as food. The modern English word for fish comes from the Old English word *fisc* (plural: *fiscas*) which was pronounced as it is today.

## Species

Over 32,000 species of fish have been described, making them the most diverse group of vertebrates. However, only a small number of species are commonly eaten by humans.

### *Mislabeling*

A recent found that one-third of the seafood sold in the U.S. had been mislabeled. The non-profit organisation Oceana had collected more than 1,200 seafood samples during 2010 and 2012. With a rate of 87 percent, snapper had been the most frequently mislabeled fish type – followed by tuna with 57 percent. Through substituting one seafood species by another can harm consumers' health. Potentially harmful toxins may enter the food supply chain and lead to foodborne illnesses.

## Common Species of Fish Used for Food

| | *Mild flavour* | *Moderate flavour* | *Full flavour* |
|---|---|---|---|
| Delicate texture | basa, flounder, hake, scup, smelt, rainbow trout, hardshell clam, blue crab, peekytoe crab, spanner crab, cuttlefish, eastern oyster, Pacific oyster | anchovy, herring, lingcod, moi, orange roughy, Atlantic ocean perch, Lake Victoria perch, yellow perch, European oyster, sea urchin | Atlantic mackerel |
| Medium texture | black sea bass, European sea bass, hybrid striped bass, bream, cod, drum, haddock, hoki, Alaska pollock, rockfish, pink salmon, snapper, tilapia, turbot, walleye, lake whitefish, wolffish, hardshell clam, surf clam, cockle, Jonah crab, snow crab, crayfish, bay scallop, Chinese white shrimp | sablefish, Atlantic salmon, coho salmon, skate, dungeness crab, king crab, blue mussel, greenshell mussel, pink shrimp | escolar, chinook salmon, chum salmon, American shad |
| Firm texture | arctic char, carp, catfish, dory, grouper, halibut, monkfish, pompano, Dover sole, sturgeon, tilefish, wahoo, yellowtail, Abalone, conch, stone crab, American lobster, spiny lobster, octopus, black tiger shrimp, freshwater shrimp, gulf shrimp, Pacific white shrimp, squid | barramundi, cusk, dogfish, kingklip, mahimahi, opah, mako shark, swordfish, albacore tuna, yellowfin tuna, geoduck clam, squat lobster, sea scallop, rock shrimp | barracuda, Chilean sea bass, cobia, croaker, eel, blue marlin, mullet, sockeye salmon, bluefin tuna |

### Persistent Organic Pollutants

If fish and shellfish inhabit polluted waters, they can accumulate other toxic chemicals, particularly fat-soluble pollutants containing chlorine or bromine, dioxins or PCBs. Fish that is to be eaten should be caught in unpolluted water. Some organisations such as SeafoodWatch, RIKILT, Environmental Defence Fund, IMARES provide information on species that do not accumulate much toxins/metals.

### Parasites

Parasites in fish are a natural occurrence and common. Though not a health concern in thoroughly cooked fish, parasites are a concern when consumers eat raw or lightly preserved fish such as sashimi, sushi, ceviche, and gravlax. The popularity of such raw fish dishes makes it important for consumers to be aware of this risk. Raw fish should be frozen to an internal temperature of –20°C (–4°F) for at least 7 days to kill parasites. It is important to be aware that home freezers may not be cold enough to kill parasites.

Traditionally, fish that live all or part of their lives in fresh water were considered unsuitable for sashimi due to the possibility of parasites. Parasitic infections from freshwater fish are a serious problem in some parts of the world, particularly Southeast Asia. Fish that spend part of their life cycle in brackish or freshwater, like salmon are a particular problem. A study in Seattle, Washington showed that 100% of wild salmon had roundworm larvae capable of infecting people. In the same study farm raised salmon did not have any roundworm larvae.

Parasite infection by raw fish is rare in the developed world (fewer than 40 cases per year in the U.S.), and involves mainly three kinds of parasites: Clonorchis sinensis (a trematode/fluke), Anisakis (a nematode/roundworm) and Diphyllobothrium (a cestode/tapeworm). Infection risk of anisakis is particularly higher in fishes which may live in a river such as salmon (*sake*) in Salmonidae or mackerel (*saba*). Such parasite infections can generally be avoided by boiling, burning, preserving in salt or vinegar, or freezing overnight. In Japan it is common to eat raw salmon and ikura, but these foods are frozen overnight prior to eating to prevent infections from parasites, particularly anisakis.

### *Fish, Meat and Vegetarians*

***Figure:*** *Hawaiian cuisine: Seared ahi and wasabi beurre blanc sauce*

### *Nutritional Content of Fish Compared to Meat 110 Grams (4 oz or .25 lb)*

| *Source* | *Calories* | *Protein* | *Carbs* | *Fat* | |
|---|---|---|---|---|---|
| fish | | 110–140 | 20–25 g | 0 g | 1–5 g |
| chicken breast | | 160 | 28 g | 0 g | 7 g |
| lamb | | 250 | 30 g | 0 g | 14 g |
| steak (beef top round) | | 210 | 36 g | 0 g | 7 g |
| steak (beef T-bone) | | 450 | 25 g | 0 g | 35 g |

Meat is animal flesh that is used as food. Most often, this means the skeletal muscle and associated fat, but it may also describe other edible organs and tissues. The term "meat" is used by the meat packing industry in a more restrictive sense—the flesh of mammalian species (pigs, cattle, etc.) raised and prepared for human consumption, to the exclusion of fish and poultry.

Vegetarians don't eat fish, and consider that fish is meat, since it is the flesh of an animal. Vegans or strict vegetarians refrain from consuming any animal products, not only meat and fish but, in contrast to ovo-lacto vegetarians, eggs, dairy products and all other animal-derived substances.

However, pescetarians eat fish and other seafood, but not mammals and birds. The Merriam-Webster dictionary dates the origin of the term "pescetarian" to 1993 and defines it to mean: "one whose diet includes fish but no meat." Pescatarians may consume fish based solely upon the idea that the fish are not factory farmed as land animals are (i.e., their problem is with the capitalist-industrial production of meat, not with the consumption of animal foods themselves). However, this is an incorrect assumption, as fish are often raised in artificial environments, with the same types of cramped, unnatural, and often unsanitary conditions that land animals are raised in. Some eat fish with the justification that fish have less sophisticated nervous systems than land-dwelling animals. Others may choose to consume only wild fish based upon the lack of confinement, while choosing to not consume fish that have been farmed.

A 1999 metastudy combined data from five studies from western countries. The metastudy reported mortality ratios, where lower numbers indicated fewer deaths, for pescetarians (fish eaters) to be 0.82, vegetarians to be 0.84, occasional meat eaters to be 0.84. Regular meat eaters and vegans shared the highest mortality ratio of 1.00. The study reported the numbers of deaths in each category, and expected error ranges for each ratio, and adjustments made to the data. However, the "lower mortality was due largely to the relatively low prevalence of smoking in these [vegetarian] cohorts".

### In Religion

Religious rites and rituals regarding food also tend to classify the birds of the air and the fish of the sea separately from land-bound mammals. Sea-bound mammals are often treated as fish under religious laws - as in Jewish dietary law, which forbids the eating of whale, dolphin, porpoise, and orca because they are not "fish with fins and scales"; nor, as mammals, do they chew their cud and have cloven hooves, as required by Leviticus 11:9-12. Jewish (kosher) practice treat fish differently from other animal foods.

***Figure:*** *Roast fish*

The distinction between fish and "meat" is codified by the Jewish dietary law of *kashrut*, regarding the mixing of milk and meat, which does not forbid the mixing of milk and fish. Modern Jewish legal practice (*halakha*) on *kashrut* classifies the flesh of both mammals and birds as "meat"; fish are considered to be *parve*, neither meat nor a dairy food. (The preceding portion refers only to the halakha of Ashkenazi Jews Sephardic Jews do not mix fish with dairy)

Seasonal religious prohibitions against eating meat do not usually include fish. For example, non-fish meat was forbidden during Lent and on all Fridays of the year in pre-Vatican II Roman Catholicism, but fish was permitted (as were eggs). In Eastern Orthodoxy, fish is permitted on some fast days when other meat is forbidden, but stricter fast days also prohibit fish with spines, while permitting invertebrate seafood such as shrimp and oysters, considering them "fish without blood."

Some Buddhists and Hindus (Brahmins of West Bengal, Odisha and Saraswat Brahmins of the Konkan) abjure meat that is not fish. Muslim (halaal) practice also treats fish differently from other animal foods, as it can be eaten.

### Taboos on Eating Fish

Among the Somali people, most clans have a taboo against the consumption of fish, and do not intermarry with the few occupational clans that do eat it. There are taboos on eating fish among many upland pastoralists and agriculturalists (and even some coastal peoples) inhabiting parts of southeastern Egypt, Ethiopia, Eritrea, Somalia, Kenya, and northern Tanzania. This is sometimes referred to as the "Cushitic fish-taboo", as Cushitic speakers are believed to have been responsible for the introduction of fish avoidance to East Africa, though not all Cushitic groups avoid fish. The zone of the fish taboo roughly coincides with the area where Cushitic languages are spoken, and as a general rule, speakers of Nilo-Saharan and Semitic languages do not have this taboo, and indeed many are watermen. The few Bantu and Nilotic groups in East Africa that do practice fish avoidance also reside in areas where Cushites appear to have lived in earlier times. Within East Africa, the fish taboo is found no further than Tanzania. This is attributed to the local presence of the tsetse fly and in areas beyond, which likely acted as a barrier to further southern migrations by wandering pastoralists, the principal fish-avoiders. Zambia and Mozambique's Bantus were therefore spared subjugation by pastoral groups, and they consequently nearly all consume fish.

There is also another centre of fish avoidance in Southern Africa, among mainly Bantu speakers. It is not clear whether this disinclination developed independently or whether it was introduced. It is certain, however, that no avoidance of fish occurs among southern Africa's earliest inhabitants, the Khoisan. Nevertheless, since the Bantu of southern Africa also share various cultural traits with the pastoralists further north in East Africa, it is believed that, at an unknown date, the taboo against the consumption of fish was similarly introduced from East Africa by cattle-herding peoples who somehow managed to get their livestock past the aforementioned tsetse fly endemic regions. Certain species of fish are also forbidden in Judaism such as the freshwater eel (Anguillidae) and all species of catfish. Although they live in water, they appear to have no fins or scales (except under a microscope). Sunni Muslim laws are more flexible in this and catfish and shark are generally seen as halal as they are special types of fish. Eel is generally considered permissible in the four Sunni *madh'hab*, but the Ja'fari jurisprudence followed by most Shia Muslims forbids it.

Many tribes of the Southwestern United States, including the Navaho, Apache, and Zuñi, have a taboo against fish and other water-related animals, including waterfowl.

### *Preparation*

Fish can be prepared in a variety of ways. It can be uncooked (raw) (*cf.* sashimi). It can be cured by marinating (*cf.* escabeche), pickling (*cf.* pickled herring), or smoking (*cf.* smoked salmon). Or it can be cooked by baking, frying (*cf.* fish and chips), grilling, poaching (*cf.* court-bouillon), or steaming. Many of the preservation techniques used in different cultures have since become unnecessary but are still performed for their resulting taste and texture when consumed.

## Dishes

### *Bokkoms*

Bokkoms (or Bokkems) is whole, salted and dried mullet (more specifically the Southern mullet, *Liza richardsonii*, a type of fish commonly known in the Western Cape of South Africa as "harders"), and is a well-known delicacy from the West Coast region of South Africa. This salted fish is dried in the sun and wind and is eaten raw after pealing off the skin. In some cases it is also smoked. It is sometimes referred to as "fish biltong".

The word *bokkom* comes from the Dutch word *bokkem*, which is a variant of the word *bokking* (or *buckinc* in Middle Dutch). The word *bokking* is derived from the word *bok* (the Dutch word for *buck* or *goat*), and refers to the fact that bokkoms reminds of goat, due to the fact that bokkoms has the same shape as the horns of a goat, it is just as hard as a goat's horns, and it stinks just as much as the horn of a goat (goats have scent glands behind their horns which causes the smell). The first official record of the use of the word in the Afrikaans language of South Africa was in the *Patriotwoordeboek* in 1902 in the form *bokkom*.

### *Origin*

As early as 1658, only six years after the first permanent settlement of Europeans at the Cape of Good Hope, four free burghers were given permission to settle in Saldanha Bay on the West Coast of Southern Africa. They were given the right from the Dutch East Indian Company to fish the waters of Saldanha Bay and send their catches to the Company's trading post at the Cape of Good Hope to be sold to other burghers as well as to passing ships in Table Bay. They had sole rights to the lucrative fishing until 1711. One fifth of the catch had to be delivered in salted and dried form, which was the origins of bokkoms in South Africa.

### Ingredients

The ingredients for bokkoms consists of small (juvenile) mullet (fish), coarse salt and fresh water.

### Preparation

The original West Coast way of preparing bokkoms starts with catching the small (juvenile) mullets (called "harders" in Afrikaans). A large square tank built out of bricks or stone is filled with a strong pickle made of around 50 kg of coarse salt and fresh water, to which the fish is added. As soon as enough fish is added to reach the top of the tank, two or three spades full of dry coarse salt is spread on top of the fish. Thereafter more layers of fish and salt is put on top until all the fish is covered in salt. A thick layer of salt is added to the top. This is left for one day. On the second day, a press made with wood with weights on top, is placed on the fish in the tank. The purpose of the press is to ensure that the guts of the fish is pressed flat so that it does not go bad. After the third day in the tank, the fish is taken out and is then stringed up in bunches of 10 to 25 fish each on a rope, making use of a fish needle which is pushed through the eyes of the fish. The bunches are then dipped 2 to 3 times in fresh water, before it is hung on scaffolds to dry.

### Drying

The most suitable circumstances for drying is a lot of wind and not too harsh sunlight. At nighttime, the fish is brought under a roof to prevent it from drawing in damp, and the next day it is taken outside again to hang in the sun. The whole drying process takes about 5 days. More recently, people have started to make use of drying "ovens". This consists of a closed room in which the fish are hung and which contains an electrical fan which blows in heated air into the room. This simplifies the drying process considerably, since the fish does not need to be moved indoors every evening, and it also allows drying to take place during the wetter winter months.

### Comparison to Dried Herring, Bloaters, Kippers and Buckling

Bokkoms are similar to dried herring, as well as bloaters, kippers and buckling. Bloaters (whole fish), kippers (split) and buckling (head and guts removed) are all lightly salted and smoked for a short period of time. Although the method for making bokkoms probably originated from the European method of making dried herring, bokkoms use a different species of fish (the mullet) and do not regularly make use of smoking in preparation of the delicacy.

### *The Bokkom Industry*

Bokkoms is a delicacy typical to the Western Cape of South Africa, and specifically the West Coast. Although it has a significant market in the Western Cape (where it is very popular and can be found in hotels, bars, bottle stores (in UK: off-licence), fish shops and beach kiosks on the beaches and holiday resorts), it has not been able to become a common product in the larger part of Southern African butcheries, fisheries and grocery stores like biltong has. It is, however, available for ordering and shipping through the internet in vacuum packed format.

There is a vibrant bokkom industry in the "bokkom capital" of the world, Velddrif, on the West Coast of South Africa. Around 95% of South Africa's bokkoms are produced in Velddrif, in a series of small individual factories located along the Berg river. Each factory has its own small jetty on the river at the front of the factory. In the past, large schools of mullet was caught in the river and the jetties were used as a place to tie the fishermen's "bakkies" (small boats) to unload their catch. Because of over-fishing, the catching of mullet in the river is now prohibited and it must be netted in the open sea just off Laaiplek.

Velddrif is the ideal place for the bokkom industry. It has access to the mullet just off the coast, its weather conditions are ideal for drying the fish (dry summers and relatively low rainfall), and it has access to huge amounts of sea salt as spring tide pushes the sea over the extensive salt pans (the largest salt factory in South Africa where salt is extracted from sea water, the Cerebos salt factory, is also situated there). It also has access to fresh water in the form of the Berg river. It is thus no co-incidence that the bokkom industry originated and still continue to be situated in Velddrif. The well-known "Bokkomlaan" (Bokkom avenue) along the banks of the Berg river in Velddrif is a place where tourists can see large numbers of bokkoms strung up in bunches on rows of reed scaffolds along the side of the road.

### *Eating*

Bokkoms is a unique, traditional delicacy of the West Coast of South Africa. It is best enjoyed with white wine, or with bread, apricot jam and black coffee, but can also be used in soups and spaghetti's, tapenades, ragout or just as a bite on its own.

### *Environmental Concerns*

There has been concern raised over declining fish stocks on the West Coast of South Africa. The Southern mullet is on the WWF

Southern African Sustainable Seafood Initiative (SASSI) Orange list as an unsustainable species. This group includes species that have associated reasons for concern, either because the species is depleted as a result of overfishing and cannot sustain current fishing pressure, or because the fishing or farming method poses harm to the environment and/or the biology of the species makes it vulnerable to high fishing pressure. These species may be legally sold by registered commercial fishers and retailers. However, an increased demand for these could compromise a sustainable supply.

### *In Popular Culture*

The word *bokkoms* has been taken up in the Afrikaans language in at least three proverbs:

1. To refer to a very boring, tiresome, dull person, i.e. "a dry old stick", in the phrase "`n droë bokkom" (literally translated "a dry bokkem").
2. To refer to someone who is incompetent, unable to perform his work adequately, not able to do the simplest of things, in the phrase "hy kan nie bokkom braai nie" (literally translated "he is not able to barbecue a bokkom").
3. To refer to someone who is very thin and emaciated, in the phrase "hy is pure bokkom en biltong, maar min vir die jakkalse" (literally translated "he is pure bokkom and biltong, but not much for the jackal").

It has also been used as the name for a specific family of plants with sharp seedpods found in South Africa. These plants have been given the name of bokkoms in Afrikaans due to the fact that the seedpods look like bokkoms, and their unpleasant smell reminds of the strong fish smell of bokkoms.

### Biltong

Biltong is a variety of cured meat that originated in South Africa. Various types of meat are used to produce it, ranging from beef and game meats to fillets of ostrich from commercial farms. It is typically made from raw fillets of meat cut into strips following the grain of the muscle, or flat pieces sliced across the grain. It is similar to beef jerky in that they are both spiced, dried meats. The typical ingredients, taste and production processes differ, the main difference being that biltong is usually thicker (from cuts up to 1" (25 mm) thick), while jerky is rarely more than 1/8" (3 mm) thick. Also, biltong does not have a sweet taste.

The word *biltong* is from the Dutch *bil* ("rump") and *tong* ("strip" or "tongue").

### Origins

The settlers (Dutch, German, French) who arrived in southern Africa in the early 17th century brought recipes for curing and drying meat from Europe. Preparation involved applying vinegar and rubbing the strips of meat with a mixture of salts (saltpetre) and spices including pepper, coriander and cloves.

The need for preservation in the new colony was pressing. Building up herds of livestock took a long time but with indigenous game in abundance, traditional methods were available to preserve large masses of meat such as found in the eland in a hot climate. Iceboxes and fridges had not been invented yet. Biltong as it is today evolved from the dried meat carried by the wagon-travelling Voortrekkers, who needed stocks of durable food as they migrated from the Cape Colony north and north-eastward (away from British rule) into the interior of Southern Africa during the Great Trek. The meat was preserved and hung to be dried for a fortnight after which it would be ready for packing in cloth bags.

### Ingredients

The most common ingredients of biltong are:

- Meat
- Black pepper
- Coriander
- Salt
- Sugar or Brown sugar
- Vinegar

Modern-day ingredients sometimes added include: balsamic vinegar or malt vinegar, dry ground chili peppers, nutmeg, garlic, bicarbonate of soda, Worcestershire sauce, onion powder, and saltpetre.

### Meat

Prior to the introduction of refrigeration, the curing process was used to preserve all kinds of meat in South Africa. However today biltong is most commonly made from beef, primarily because of its widespread availability and lower cost relative to game. For the finest cuts, fillet, sirloin or steaks cut from the hip such as topside or silverside. Other cuts can be used, but are not as high in quality.

Biltong can also be made from:

- Chicken, simply referred to as 'chicken biltong'
- Fish in this case, known as *bokkoms* (shark biltong can also be found in South Africa).
- Game such as kudu and springbok
- Ostrich meat (bright red, often resembling game)

*Bokkoms* should not be confused with other cured fish such as dried angelfish and dried snoek.

### Preparation

Ideally the meat is marinated in a vinegar solution (grape vinegar is traditional but balsamic and cider also works very well) for a few hours, this being finally poured off before the meat is flavoured.

The spice mix traditionally consists equal amounts of: rock salt, whole coriander slightly roasted and roughly ground, black pepper and brown sugar. This mix is then ground roughly together, sprinkled liberally over the meat and rubbed in. Saltpetre is optional and can be added as an extra preservative (necessary only for wet biltong that is not going to be frozen).

The meat should then be left for a further few hours (or refrigerated overnight) and any excess liquid poured off before the meat is hung in the dryer.

### Drying

**Figure:** *Biltong quick drying using an electric oven*

It is typically dried out in the cold air (rural settings), cardboard or wooden boxes (urban) or climate-controlled dry rooms (commercial). Depending on the spices used, a variety of flavours may be produced. Biltong can also be made in colder climates by using an electric lamp to dry the meat, but care must be taken to ventilate, as mould can begin to form on the meat.

A traditional slow dry will deliver a medium cure in about 4 days.

An electric fan-assisted oven set to 40–70 °C (100–160 °F), with the door open a fraction to let out moist air, can dry the meat in approximately 4 hours. Although slow dried meat is considered by some to taste better, oven dried is ready to eat a day or two after preparation.

### Comparison to Jerky

Biltong differs from jerky in two distinct ways:

- The meat used in biltong can be much thicker; typically biltong meat is cut in strips approx 1" (25 mm) wide – but can be thicker. Jerky is normally very thin meat.
- The vinegar, salt and spices in biltong, together with the drying process, cures the meat as well as adding texture and flavour. Jerky is traditionally dried with salt but without vinegar.

### Retail

Biltong is a common product in Southern African butcheries and grocery stores, and can be bought in the form of wide strips (known as *stokkies*, meaning "little sticks"). It is also sold in plastic bags, sometimes shrink-wrapped, and may be either finely shredded or sliced as biltong chips.

There are also specialised retailers that sell biltong. These shops may sell biltong as "wet" (moist), "medium" or "dry". Additionally, some customers prefer it with a lot of fat within the muscle fibres, while others prefer it as lean as possible.

A 2013 study analysed samples of game meat from South African supermarkets, wholesalers, and other outlets. It was found that some types of biltong labelled "kudu", "springbok", or "ostrich" were made of hartebeest. Of 146 samples, 100 were mislabelled, which revealed a great problem in meat labelling in South Africa.

### Eating

While biltong is usually eaten as a snack, it can also be diced up into stews, added to muffins and pot bread. Biltong-flavoured potato crisps have also been produced. Biltong can be used as a teething aid for babies. Some retail stores offer a mild form of biltong especially for this purpose which does not contain the spices used for flavouring.

### *Biltong Worldwide*

Biltong's popularity has spread to many other countries, notably Canada, the United Kingdom, Australia, New Zealand and now India which have large South African populations, and also to the United States. Biltong is also produced within South African expatriate communities across the globe, for example in Germany and even South Korea. Biltong produced in South Africa may not be imported into Britain, according to rules governing the importation of meat-based products from non-EU countries laid down by Customs & Excise department, thus it is made in the UK.

## Bouillabaisse

Bouillabaisse is a traditional Provençal fish stew originating from the port city of Marseille. The French and English form *bouillabaisse* comes from the Provençal Occitan word *bolhabaissa*, a compound that consists of the two verbs *bolhir* (to boil) and *abaissar* (to reduce heat, i.e., simmer).

Bouillabaisse originally was a stew made by Marseille fishermen using the bony rockfish which they were unable to sell to restaurants or markets. There are at least three kinds of fish in a traditional bouillabaisse: typically red rascasse (Scorpaena scrofa); sea robin (fr: *grondin*); and European conger (fr: *congre*). It can also include gilt-head bream (fr: *dorade*); turbot; monkfish (fr: *lotte* or *baudroie*); mullet; or silver hake (fr: *merlan*). It usually also includes shellfish and other seafood such as sea urchins (fr: *oursins*), mussels (fr: *moules*); velvet crabs (fr: *étrilles*); spider crab (fr: *araignées de mer*) or octopus. More expensive versions may add langoustine (European lobster), though this was not part of the traditional dish made by Marseille fishermen. Vegetables such as leeks, onions, tomatoes, celery and potatoes are simmered together with the broth and served with the fish. The broth is traditionally served with a rouille, a mayonnaise made of olive oil, garlic, saffron and cayenne pepper on grilled slices of bread.

What makes a bouillabaisse different from other fish soups is the selection of Provençal herbs and spices in the broth; the use of bony local Mediterranean fish; the way the fish are added one at a time, in a certain order, and brought to a boil; and the method of serving. In Marseille, the broth is served first in a bowl containing the bread and rouille, with the seafood and vegetables served separately in another bowl or on a platter.

### *Marseille Bouillabaisse*

Recipes for bouillabaisse vary from family to family in Marseille, and local restaurants dispute which versions are the most authentic. In Marseille, bouillabaisse is rarely made for fewer than ten people; the more people who share the meal, and the more different fish that are included, the better the bouillabaisse.

An authentic Marseille bouillabaisse must include *rascasse* (Scorpaena scrofa), a bony rockfish which lives in the calanque and reefs close to shore. It usually also has *congre* (eng: European conger) and *grondin* (eng: sea robin). According to the *Michelin Guide Vert*, the four essential elements of a true bouillabaisse are the presence of rascasse, the freshness of the fish; olive oil, and an excellent saffron.

The American chef and food writer Julia Child, who lived in Marseille for a year, wrote: "to me the telling flavour of bouillabaisse comes from two things: the Provençal soup base — garlic, onions, tomatoes, olive oil, fennel, saffron, thyme, bay, and usually a bit of dried orange peel — and, of course, the fish — lean (non-oily), firm-fleshed, soft-fleshed, gelatinous, and shellfish."

This is the recipe from one of the most traditional Marseille restaurants, Grand Bar des Goudes on Rue Désirée-Pelleprat:

- 4 kilograms of fish and shellfish:
    - o *grondin* (eng. sea robin)
    - o *Rascasse* (Scorpaena scrofa);
    - o *rouget grondin* (red gurnard);
    - o *congre* (eng. conger);
    - o *baudroie* (lotte, or monkfish);
    - o *Saint-Pierre* (eng. John Dory);
    - o *vive* (eng. Weever);
    - o Octopus
    - o 10 sea urchins
- 1 kilogram of potatoes
- 7 cloves of garlic
- 3 onions
- 5 ripe tomatoes
- 1 cup of olive oil

- 1 bouquet garni
- 1 branch of fennel
- 8 pistils of saffron
- 10 slices of pain de campagne (country bread)
- salt and Cayenne pepper

### *The Rouille*

- 1 egg yolk
- 2 cloves of garlic
- 1 cup of olive oil
- 10 pistils of saffron
- salt and Cayenne pepper

1. Clean and scale the fish and wash them, if possible in sea water. Cut them into large slices, leaving the bones. Wash the octopus and cut into pieces.
2. Put the olive oil in a large casserole. Add the onions, cleaned and sliced; 6 cloves of garlic, crushed; the pieces of octopus, and the tomatoes peeled and quartered, without seeds. Brown at low heat, turning gently for five minutes, for the oil to take in the flavours.
3. Add the sliced fish, beginning with the thickest to the smallest. Cover with boiling water, and add the salt and the pepper, the fennel, the bouquet garni and the saffron. Boil at a low heat, stirring from time to time so the fish doesn't stick to the casserole. Correct the seasoning. The bouillabaisse is cooked when the juice of the cooking is well blended with the oil and the water. (about twenty minutes).
4. Prepare the rouille: Remove the stem of the garlic, crush the cloves into a fine paste with a pestle in a mortar. Add the egg yolk and the saffron, then blend in the olive oil little by little to make a mayonnaise, stirring it with the pestle.
5. Cook the potatoes, peeled and boiled and cut into large slices, in salted water for 15 to 20 minutes. Open the sea urchins with a pair of scissors and remove the *Corail* with a small spoon.
6. Arrange the fish on a platter. Add the *corail* of the sea urchins into the broth and stir.

Serve the bouillon very hot with the rouille in bowls over thick slices of bread rubbed with garlic. Then serve the fish and the potatoes on a separate platter. Another version of the classic Marseille bouillabaisse, presented in the *Petit LaRousse de la Cuisine*, uses congre, dorade, grondin, lotte, merlan, rascasse, saint-pierre, and velvet crabs (étrilles), and includes leeks. In this version, the heads and trimmings of the fish are put together with onions, celery and garlic browned in olive oil, and covered with boiling water for twenty minutes. Then the vegetables and bouquet garni are added, and then the pieces of fish in a specific order; first the rascasse, then the grondin, the lotte, congre, dorade, etrilles, and saffran. The dish is cooked for eight minutes over high heat. Then the most delicate fish, the saint pierre and merlan, are added, and the dish is cooked another 5–8 minutes. The broth is then served over bread with the rouille on top, and the fish and crabs are served on a large platter.

Other variations add different seasonings, such as orange peel, and sometimes a cup of white wine or cognac is added.

### The Bouillabaisse of Marcel Pagnol

The French screenwriter and playwright Marcel Pagnol, a member of the Academie Francaise and a native of Marseille, showed his own idea of a proper bouillabaisse in two of his films. In *Cigalon (fr)* (1935), the chef Cigalon serves a *bouillabaisse provencale aux poisson du roche*, (Bouillabaisse of Provence with rockfish) made with a kilogram of local fish; Scorpaena scrofa (*rascasse*); Capelin; Angler fish (*baudroie*); John Dory (*Saint-Pierre*); and Slipper lobster (*Cigale de mer*). "When I put these fish into the pan," Cigalon says, "they were still wiggling their tails." Cigalons specifies that the slices of bread served with the broth should be thick and not toasted, and that the rouille "should not have too much pepper."

In the 1936 film *César (film)*, Pagnol's hero Marius reveals the secret of the bouillabaisse of a small bistro near the port in Marseille. "Everbody knows it," Marius says: "they perfume the broth with a cream of sea urchins."

### History and Legend

According to tradition, the origins of the dish date back to the time of the Phoceans, an Ancient Greek people who founded Marseille in 600 BC. Then, the population ate a simple fish stew known in Greek as 'kakavia.' Something similar to Bouillabaisse also appears in Roman mythology: it is the soup that Venus fed to Vulcan.

The dish known today as bouillabaisse was created by Marseille fishermen who wanted to make a meal when they returned to port. Rather than using the more expensive fish, they cooked the common rockfish and shellfish that they pulled up with their nets and lines, usually fish that were too bony to serve in restaurants, cooking them in a cauldron of sea water on a wood fire and seasoning them with garlic and fennel. Tomatoes were added to the recipe in the 17th century, after their introduction from America.

In the 19th century, as Marseille became more prosperous, restaurants and hotels began to serve bouillabaisse to upper-class patrons. The recipe of bouillabaisse became more refined, with the substitution of fish stock for boiling water, and the addition of saffron. Bouillabaisse spread from Marseille to Paris, and then gradually around the world, adapted to local ingredients and tastes.

The name bouillabaisse comes from the method of the preparation — the ingredients are not added all at once. The broth is first boiled (*bolh*) then the different kinds of fish are added one by one, and each time the broth comes to a boil, the heat is lowered (*abaissa*).

Generally similar dishes are found in Portugal (caldeirada), Spain (sopa de pescado y marisco, suquet de peix (es)), Italy (zuppa di pesce), Greece and all the countries bordering the Mediterranean Sea; where these kind of dishes have been made since the Neolithic Era. What makes a bouilabaisse different from these other dishes are the local Provençal herbs and spices, the particular selection of bony Mediterranean coastal fish and the way the broth is served separately from the fish and vegetables.

## Bourdeto

Bourdeto is a dish from Corfu. It comes from the Venetian word *brodeto* which means broth. It is fish cooked in a tomato sauce with onion, garlic and red spicy pepper. The best fish for bourdeto is scorpion fish. One can also find the same dish containing fillet of a bigger kind of fish. A similar dish found on the western Adriatic coast is called Brudet.

## Ceviche

Ceviche; also spelled *cebiche*, or *seviche*) is a seafood dish popular in the coastal regions of the Americas, especially Central and South America. The dish is typically made from fresh raw fish marinated in citrus juices, such as lemon or lime, and spiced with *ají* or chili peppers. Additional seasonings, such as chopped onions, salt, and coriander, may also be added. Ceviche is usually accompanied by side dishes that complement its flavours, such as sweet potato, lettuce, corn, avocado or plantain. As the dish is not cooked with heat, it must be prepared fresh to minimize the risk of food poisoning. It may be safer to prepare it with frozen or blast-frozen fish due to *Anisakis* parasites.

The origin of ceviche is disputed. Possible origin sites for the dish include the western coast of north-central South America, or in Central America. The invention of the dish is also attributed to other coastal societies, such as the Polynesian islands of the south Pacific. The Spanish, who brought from Europe citrus fruits, such as lime, could have also originated the dish with roots in Moorish cuisine. However, the most likely origin lies in the area of present-day Peru.

Along with an archaeological record suggesting the consumption of a food similar to ceviche nearly 2,000 years ago, historians believe the predecessor to the dish was brought to Peru by Moorish women from Granada, who accompanied the Spanish conquistadors and colonizers, and this dish eventually evolved into what now is considered ceviche. Peruvian chef Gastón Acurio further explains the dominant position that Lima held through four centuries as the capital of the Viceroyalty of Peru allowed for popular plates such as ceviche to be brought to other Spanish colonies in the region, and in time they became a part of local cuisine by incorporating regional flavours and styles.

Ceviche is nowadays a popular international dish prepared in a variety of ways throughout the Americas, reaching the United States in the 1980s. The greatest variety of ceviches are found in Peru; but other distinctly unique styles can also be found in coastal El Salvador, Guatemala, the United States, Mexico, Panama, the Caribbean, and several other nations.

In regard to its origin, various explanations are given. According to some historic sources from Peru, ceviche would have originated among the Moche, a coastal civilization that began to flourish in the area of current-day northern Peru nearly 2000 years ago. The Moche apparently used the fermented juice from the local *banana passionfruit*. Recent investigations further show, during the Inca Empire, fish were marinated with the use of *chicha*, an Andean fermented beverage. Different chronicles also report, along the Peruvian coast prior to the arrival of Europeans, fish was consumed with salt and *ají*. Furthermore, this theory proposes the natives simply switched to the citrus fruits brought by the Spanish colonists, but the main concepts of the plate remain essentially the same.

The invention of the dish is also attributed to places ranging from Central America to the Polynesian islands in the South Pacific. In Ecuador, it could have also had its origins with its coastal civilizations, as both Peru and Ecuador have shared cultural heritages (such as the Inca empire) and a large variety of fish and shellfish. Ceviche is not native to Mexico, despite the fact that the dish has been a part of traditional Mexican coastal cuisine for centuries. The Spanish, who brought from Europe citrus fruits such as lime, could have originated the dish in Spain with roots in Moorish cuisine.

Nevertheless, most historians agree ceviche originated during colonial times in the area of present-day Peru. They propose the predecessor to the dish was brought to Peru by Moorish women from Granada who accompanied the Spaniards, and this dish eventually evolved into what nowadays is considered ceviche. Peruvian chef Gastón Acurio further explains the dominant position that Lima held through four centuries as the capital of the Viceroyalty of Peru allowed for popular dishes such as ceviche to be brought to other Spanish colonies in the region, and in time they became a part of local cuisine by incorporating regional flavours and styles. Other notable chefs who support the Peruvian origin of the plate include Chilean Christopher Carpentier and Spaniard Ferran Adrià, who in an interview stated, "Cebiche was born in Peru, and so the authentic and genuine [cebiche] is Peruvian."

### Spread

***Preparation and Variants:*** Ceviche is marinated in a citrus-based mixture, with lemons and limes being the most commonly used. In addition to adding flavour, the citric acid causes the proteins in the seafood to become denatured, appearing to be cooked. (However, acid marinades will not kill bacteria or parasitic worms, unlike the heat of cooking.) Traditional-style ceviche was marinated for about three hours. Modern-style ceviche, popularized in the 1970s, usually has a very short marinating period. With the appropriate fish, it can marinate in the time it takes to mix the ingredients, serve, and carry the ceviche to the table.

Most Latin American countries have given ceviche its own touch of individuality by adding their own particular garnishes.

### South America

In Peru, ceviche has been declared to be part of Peru's "national heritage" and has even had a holiday declared in its honour. The classic Peruvian ceviche is composed of chunks of raw fish, marinated in freshly squeezed key lime or bitter orange (*naranja agria*) juice, with sliced onions, chili peppers, salt and pepper. Corvina or cebo (sea bass) was the fish traditionally used. The mixture was traditionally marinated for several hours and served at room temperature, with chunks of corn-on-the-cob, and slices of cooked sweet potato. Regional or contemporary variations include garlic, fish bone broth, minced Peruvian *ají limo*, or the Andean chili *rocoto*, toasted corn or *cancha* and *yuyo* (seaweed). A speciality of Trujillo is ceviche prepared from shark (*tollo* or tojo). *Lenguado* (sole) is often used in Lima. The modern version of Peruvian ceviche, which is similar to the method used in making Japanese sashimi, consists of fish marinated for a few minutes and served promptly. It was developed in the 1970s by Peruvian-Japanese chefs including Dario Matsufuji and Humberto Sato. Many Peruvian *cevicherías* serve a small glass of the marinade (as an appetizer) along with the fish, which is called *leche de tigre* or *leche de pantera*.

In Ecuador, it is spelled as cebiche. The shrimp ceviche is sometimes made with tomato sauce for a tangy taste. The Manabí style, made with lime juice, salt and the juice provided by the cooked shrimp itself is very popular. Occasionally, ceviche is made with various types of local shellfish, such as black clam (cooked or raw), oysters (cooked or raw), spondylus (raw), barnacles (cooked percebes), among others mostly cooked. It is served in a bowl with toasted corn kernels as a side dish (fried green plantain chunks called "patacones", thinly sliced plantain chips called *chifle*, and popcorn are also typical ceviche side dishes). Well cooked Sea bass (corvina), octopus, and crab ceviches are also common in Ecuador. In all ceviches, lime juice and salt are ubiquitous ingredients.

In Chile, ceviche is often made with fillets of halibut or Patagonian toothfish, and marinated in lime and grapefruit juices, as well as finely minced garlic and red chili peppers and often fresh mint and cilantro are added.

### *Central America and the Caribbean*

In Mexico and Central America, it is served in cocktail cups with tostadas, or as a tostada topping and taco filling. Shrimp, octopus, squid, tuna, and mackerel are popular bases for Mexican ceviche. The marinade ingredients include salt, lime, onion, chili peppers, avocado, and coriander leaves (known as cilantro in the Americas). Tomatoes are often added to the preparation. According to the book "Mexico One

Plate At A Time," even though the dish has been a part of traditional Mexican coastal cuisine for centuries, ceviche is not a dish native to Mexico. Despite this, Mexican ceviche has developed its own distinct styles that make it unique from other available variations.

In El Salvador the ceviche tradition is very strong. One of the most exotic ceviche recipes is "Ceviche de Concha Negra,", known in Mexico as Pata de Mula, or "The Black Clam." It is dark, nearly black, with a distinct look and flavour. It is prepared with lime juice, onion, yerba buena, salt, pepper, tomato, Worcester sauce, and sometimes picante (any kind of hot sauce or any kind of hot pepper) as desired.

In Costa Rica, the dish includes marinated fish, lime juice, salt, ground black pepper, finely minced onions, cilantro and finely minced peppers. It is usually served in a cocktail glass with a lettuce leaf and soda crackers on the side, as in Mexico. Popular condiments are tomato ketchup and tabasco sauce. The fish is typically tilapia or corvina, although mahi-mahi, shark and marlin are also popular.

In Panama, ceviche is prepared with lemon juice, chopped onion, celery, habanero pepper, and sea salt. Ceviche de corvina (white sea bass) is very popular and is served as an appetizer in most local restaurants. It is also commonly prepared with octopus, shrimp, and squid, or served with little pastry shells called "canastitas."

In Cuba, ceviche is often made using mahi-mahi prepared with lime juice, salt, onion, green pepper, habanero pepper, and a touch of allspice. Squid and tuna are also popular. In Puerto Rico and other places in the Caribbean, the dish is prepared with coconut milk. In The Bahamas and south Florida, a conch ceviche known as 'conch salad' is very popular. It is prepared by marinating diced fresh conch in lime with chopped onions, celery, and bell pepper. Diced pequin pepper and/or scotch bonnet pepper is often added for spice. In south Florida, it is common to encounter a variation to which tomato juice has been added.

### *Health Risks*

Aside from contaminants, raw seafood can also be the vector for various pathogens, viral, bacterial, as well as larger parasitic creatures. According to the 2009 Food Code published by the United States Food and Drug Administration (FDA), specific microbial hazards in ceviche include: *Anisakis simplex, Diphyllobothrium spp., Pseudoterranova decipiens,* and *Vibrio parahaemolyticus.* Anisakiasis is a zoonotic disease caused by the ingestion of larval nematodes in raw seafood dishes such as ceviche. The Latin American cholera outbreaks in the

1990s have been attributed to the consumption of raw cholera-infested seafood that was eaten as ceviche. Other studies concluded that the lack of sanitary food supply conditions, including "unwashed fruit and vegetables, contaminated food and ice from street vendors, contaminated drinking water, and contaminated crab meat transported in luggage" caused the epidemic.

## Cioppino

Cioppino is a fish stew originating in San Francisco. It is considered an Italian-American dish, and is related to various regional fish soups and stews of Italian cuisine. Cioppino is traditionally made from the catch of the day, which in the dish's place of origin is typically a combination of dungeness crab, clams, shrimp, scallops, squid, mussels and fish. The seafood is then combined with fresh tomatoes in a wine sauce, and served with toasted bread, either sourdough or baguette. The dish is comparable to cacciucco and brodetto from Italy, as well as other fish dishes from the Mediterranean region such as bouillabaisse, burrida, and bourride of the French Provence, and suquet de peix from Catalan speaking regions of coastal Spain.

### *History*

Cioppino was developed in the late 1800s by Italian fishermen who settled in the North Beach section of San Francisco, many from

Genoa, Italy. Originally it was made on the boats while out at sea and later became a staple as Italian restaurants proliferated in San Francisco.

The name comes from *ciuppin*, a word in the Ligurian dialect of the port city of Genoa, meaning "to chop" or "chopped" which described the process of making the stew by chopping up various leftovers of the day's catch. *Ciuppin* is also a classic soup of Genoa, similar in flavour to cioppino, with less tomato, and the seafood cooked to the point that it falls apart.

### Presentation

Generally the seafood is cooked in broth and served in the shell, including the crab (if any) that is often served halved or quartered. It therefore requires special utensils, typically a crab fork and cracker. Depending on the restaurant, it may be accompanied by a bib to prevent food stains on clothing (sometimes encouraged by restaurants for patrons to use as a sign to attract attention to the restaurant's food), a damp napkin, or a second bowl for the shells. A variation, the "lazy man's" cioppino, is served with seafood shelled and crab legs cracked.

## Crab Stick

Crab sticks (imitation crab meat, seafood sticks, krab) are a form of kamaboko, a processed seafood made of finely pulverized white fish flesh (surimi), shaped and cured to resemble leg meat of snow crab or Japanese spider crab.

Crab flakes use the same mixture to form flakes instead of sticks to resemble crab meat or lobster meat.

*Sugiyo Co., Ltd.* (¹0®0è0 *Sugiyo*) of Japan first produced and patented crab sticks in 1973, as Kanikama. In 1976, The Berelson Company of San Francisco, CA USA, working with Sugiyo, introduced them internationally. Kanikama is still their common name in Japan, but internationally they are marketed under names including Krab Sticks, Ocean Sticks, Sea Legs and Imitation Crab Sticks. Legal restrictions now prevent them from being marketed as "Crab Sticks" in many places, as they usually do not have crab meat.

Alaska pollock from the North Pacific is commonly the main ingredient, often mixed with egg white (albumen) or other binding ingredient, such as the enzyme transglutaminase. Crab flavouring is added (either artificial or crab-derived), and a layer of red food colouring is applied to the outside.

Individual sticks in many forms are coloured red or yellowish red on the outside, with a rectangular cross-section or a cross-section fiber showing white inside meat. In some forms, strings can be torn from the stick in a similar manner to string cheese. The texture is rubbery, with a salty taste and smell similar to steamed crab. Cross fiber is manufactured to be similar to the texture and fiber of snow crab or king crab legs.

Quality imitation crab is usually lower in cholesterol than real crab meat, but is also highly processed.

### *Uses in Cuisine*

A California roll is a sushi roll made with imitation crab meat, avocado, and cucumber (sometimes) rolled with sesame seeds on the outside. Russian, American, and European deli counters have salads prepared with imitation crab meat, eggs, vegetables and herbs chopped together and seasoned with mayonnaise.

## Crappit Heid

Crappit heid is a traditional Scots fish course. In Gaelic it is known as *ceann-cropaig*.

Its origins can be traced to the fishing communities of the North, Hebrides and North-Eastern Scotland in the eighteenth century. In a time when money was scarce, the more expensive fillets of fish, such as cod or haddock would be sold to market but the offal and less attractive parts were retained by the fisherfolk for the pot.

Crappit heid was a favourite midday or evening meal amongst those communities and consisted of the head of a large cod or similar sized fish, washed, descaled and then stuffed with a mixture of oats, suet, onion, white pepper and the liver of the fish in question. This was then sewn or skewered to close the aperture and boiled in seawater. The cooked dish would then be served with potatoes or other root vegetables in season.

Later variations include exchanging the seawater for a court bouillon of fish stock and onion. The resulting poaching liquid is often eaten as a soup before having the fish head.

Although once a very common dish, crappit heid has, like many traditional dishes, become a rarity. Cod livers are now harder to obtain and usually only available if the fish has been caught by local line fishermen. A healthy and nutritious dish, it is rich in carbohydrates, proteins, fats and more importantly cod liver oil.

## Croquette

A croquette is a small breadcrumbed fried food roll containing, usually as main ingredients, mashed potatoes and/or ground meat (veal, beef, chicken, or turkey), shellfish, fish, cheese, vegetables and mixed with béchamel or brown sauce, and soaked white bread, egg, onion, spices and herbs, wine, milk, beer or any of the combination thereof, sometimes with a filling, e.g. sauteed onions or mushrooms, boiled eggs (Scotch eggs).

The croquette is usually shaped into a cylinder, disk or oval shape and then deep-fried. The croquette (from the French *croquer*, "to crunch") gained worldwide popularity, both as a delicacy and as a fast food.

### *In Various Countries*

***Figure:*** *Korokke Soba*

### *Belgium*

Mashed potato filled 'kroketten/croquettes' are often served as a side dish in winter holiday meals, such as Christmas. In fast food cuisine, varieties exist without potatoes but with cheese, beef or goulash, often in a filling based on béchamel sauce. The Dutch version of little balls, bitterballen, is also found in snackbars. A typical Belgian variety is filled with grey shrimp - 'garnaalkroketten/croquettes aux crevettes'. They are typically garnished with a slice of lemon and some deep fried parsley.

### Brazil

Croquettes, primarily made from beef, are common in many parts of Brazil.

### Germany, Austria and Switzerland

Plain potato croquettes are served as side dish in restaurants and are also available frozen in supermarkets. They are usually called *Kroketten* (singular *Krokette*, fem.).

### Hungary

"Krokett" is a small cylindrical croquette similar to the Czech variety: potatoes, eggs, flour, butter, seasoned with nutmeg and salt and deep fried in oil. This variety can be ordered in most restaurants as a side dish, and also bought frozen. When made with cottage cheese they are called túrókrokett.

### India

A potato-filled croquette called *Aloo tikki* is very popular in northern India and is typically served with a stew. They are mostly eaten as snacks at home and are also popularly sold by road-side vendors. In the Indian state of West Bengal, it is called *alu chop* as in Bangladesh. Sometimes it is called a "cutlet" and eaten plain or as a fast food variation that is served inside a hamburger bun (like a vegetarian burger). Meat croquettes called *kababs* are made with minced mutton. Lightly spiced beef croquettes are a popular snack and appetiser among the Christian communities in Goa and Kerala.

### Indonesia

The kroket (Dutch) made of potato and minced chicken inside a crepe-like wrapper is one of the more popular snack items in Indonesia introduced during the Dutch colonial rule. The kroket is made by taking a potato and chicken filling and wrapping it inside a crepe-like wrapper, breaded, and fried.

### Italy

In Italy, *crocchette* (named in the south as crocchè) are made mainly with crushed potatoes or vegetables, like aubergines (*crocchette di melanzane*). They originate from Naples and Sicily, and are served as antipasti in the pizzerie all around Italy.

### Japan

A relative of the croquette, known as korokke ( ³0í0Ã0±0 ) is a very popular fried food, widely available in supermarkets and butcher shops, as well as from speciality *korokke* shops. Generally patty-shaped, it is mainly made of potatoes with some other ingredients such as vegetables (e.g. onions and carrots) and maybe less than 5% meat (e.g. pork or beef). It is often served with tonkatsu (h0"0K0d0) sauce. Cylindrically-shaped *korokke* are also served, which more closely resemble the French version, where seafood (prawns or crab meat) or chicken in white sauce (ragout) is cooled down to make it harden before the croquette is breaded and deep-fried. When it is served hot, the inside melts. This version is called "cream *korokke*" to distinguish it from the potato-based variety. It is often served with no sauce or tomato sauce. Unlike its Dutch cousin, croquettes made mainly of meat are not called *korokke* in Japan. They are called *menchi katsu* (á0ó0Á0«0Ä0), short for minced meat cutlets.

### Korea

Usually called Goroke (à¬\¸Ï) or Keuroket (lÐ\¸–Ï), it is a food that is sold in most bread shops in Korea. The most common type of Goroke are deep fried rolls stuffed with Japchae (¡ÇDÌ) ingredients or chicken curry. There are also Goroke filled with Kimchi, pork, and sometimes Bulgogi ingredients. Many Koreans stores often advertise the Goroke as a French product and is usually sold in most European style bread stores all over Korea.

### Mexico

Croquettes are usually made of tuna or chicken and potatoes.

### Netherlands

***Figure:*** *A "kalfsvleeskroket": Dutch croquette containing a veal ragout*

The ragout-filled dish was regarded as a French cuisine delicacy, first described in a recipe from 1691 by the chef of the French king Louis XIV and utilising ingredients such as truffles, sweetbread and cream cheese. From the 1800s onwards, it became a way to use up leftover stewed meat. After World War II, several suppliers started mass-producing croquettes filled with beef. The croquette subsequently became even more popular as a fast food; meat ragout covered in breadcrumbs which is subsequently deep-fried. Its success as a fast food garnered its reputation as a cheap dish of dubious quality, to such an extent that Dutch tongue in cheek urban myths relate its "allegedly mysterious content" to offal and butchering waste. Research in 2008 showed that 350 million *kroketten* are eaten in the Netherlands. It is estimated that 75% of all Dutch people will eat them, resulting in 29 *kroketten* per person per year on average. The major consumers are between 35 and 49 years old.

The success of the croquette led to a whole series of food products resembling the croquette, but with other types of fillings such as noodles, rice and kidney, and with names like *bamibal*, *nasibal* and *nierbroodje* instead of croquette. Variants of the croquette which specify the kind of meat can also be found, like *rundvleeskroket* (made with beef) and *kalfsvleeskroket* (made with veal). Also popular in Dutch snack bars is the *satékroket* where the filling consists of a peanut satay sauce and shredded meat in a ragout. A smaller round version of the standard beef or veal croquette, the bitterbal, is often served with mustard as a snack in bars and at receptions. Potato croquettes and potato balls (similar to potato croquettes, but small and round) can be bought frozen in most food stores.

Broodje kroket, a croquette on a bread roll, is sold in restaurants, snack-shops, and even by street vendors.

### Poland

Croquettes in Poland are basically made from a thin rolled pancake stuffed with mushrooms, meat, cabbage, sauerkraut or combinations of those ingredients, then covered in breadcrumbs, fried in a pan and usually served with a clear soup like borscht.

### Portugal

Croquetes are cylindrical, covered in breadcrumbs, and deep-fried. They are usually made with white sauce and beef, sometimes mixed with varying amounts of pork, and frequently with some chouriço, black pepper or piri-piri to add more flavour. Seafood, fish (other than codfish) and vegetarian (potato) croquetes are also eaten in Portugal, but those have other names (such as rissol, pastel, empada) and thus the name croquete only refers to the Dutch-style beef croquette.

### Russia

The widespread (from French *cotelette*) is made of minced meat (beef or pork or mixture of both, chicken, turkey or fish), bread, eggs, white onions, salt and spices, shaped as a meat patty and pan fried. Bread is added in amount up to 25% of meat, adding softness to the final product and also making it cheaper to produce. Another popular variation similar to French *cotelettes de volaille* is Chicken Kiev, made from boned chicken breast pounded and rolled around cold unsalted butter, then breaded and fried.

### Spain

Croquetas (the Spanish name for Croquettes), especially filled with jamón, chicken or cod, are also a typical tapas dish. Unfilled bechamel croquettes are also consumed in parts of Spain.

### United Kingdom

Plain potato croquettes are available frozen or refrigerated in most supermarkets.

### United States

"Boardwalk" fish cakes and crab cakes, eaten on the east coast of the United States, are essentially croquettes. They consist, respectively, of chopped fish or crab meat, mixed in a buttery dough which is then breaded and deep-fried.

A deviled crab (*croqueta de jaiba*) is a particular variety of a blue crab croquette from Tampa, Florida. The crab meat is seasoned with a unique Cuban-style enchilada or *sofrito* sauce (locally known as "*chilau*"), breaded with stale Cuban bread crumbs, formed into the approximate shape of a prolate spheroid, and fried. It is meant to be eaten with one hand. It originated in the immigrant community of Ybor City during a cigar workers strike in the 1920s and is still very popular in the area.

Other types of crab and fish croquettes are made throughout the southern U.S., especially in coastal states such as Louisiana where seafood dishes are common staples. These varieties have different seasonings and shapes, with some served inside the scooped-out shell of the crab.

A traditional New England/Northeastern United States preparation uses ham, usually of the maple-cured variety, along with pre-cooked mashed potato (often mixed with some mild seasonings and a bit of milk or butter for a smoother consistency) for the outer roll. These are dipped in crumbed breading, and sautéed or fried in a small skillet using butter. Typically, these are most common during the Thanksgiving-to-Christmas holiday season as one of several ways to use up leftover holiday ham.

Another croquette dish popular across much of the American South is salmon croquettes. Any canned fish - usually salmon or mackerel, although canned tuna is also used in some recipes (although the dish is often colloquially referred to as "salmon croquettes" or "salmon patties" regardless of the actual fish used) - is mashed by hand to break up any fish bones and give the fish meat a smoother consistency, then combined with a binder and various seasonings. Seasonings typically include pepper, salt, chopped (sometimes sautéed) onions, garlic, lemon juice, and/or paprika. The binder can be any starch such as flour, cornmeal, matzo meal, ground crackers of any type, even white rice or oatmeal - although these latter variations are not as common, and are mostly limited to the northern U.S. Chopped eggs, parsley, and Parmesan cheese may also be added. The mixture is then shaped into rounded patties for pan- or deep-frying; corn or peanut oil are the most commonly used frying oils in the southern U.S., but canola, safflower, or olive oil are also used, and some recipes call specifically for pan-frying in butter or margarine.

## Curanto

Curanto is a traditional food of Chiloé Archipelago that has spread to the southern areas of Chile. It is traditionally prepared in a hole, about a metre and a half (approx. one and a half yards) deep, which is dug in the ground. The bottom is covered with stones, heated in a bonfire until red.

### *Preparation*

The ingredients consist of shellfish, meat, potatoes, milcao (a kind of potato bread), chapaleles, and vegetables (sometimes including also specific types of fish). The varieties of shellfish vary but almejas (clams), cholgas (ribbed mussels) and picorocos (giant barnacles) are essential. The quantities are not fixed; the idea is that there should be a little of everything. Each layer of ingredients is covered with nalca (Chilean rhubarb) leaves, or in their absence, with fig leaves or white cabbage leaves. All this is covered with wet sacks, and then with dirt and grass chunks, creating the effect of a giant pressure cooker in which the food cooks for approximately one hour.

Curanto can also be prepared in a large stew pot that is heated over a bonfire or grill or in a pressure cooker. This *stewed curanto* is called *pulmay* in the central region of Chile.

### History

It is believed that this form of preparing foods was native to the "chono" countryside and that, with the arrival of the southern peoples and the Spanish conquistadors, new ingredients were added until it came to be the curanto that is known today. However, there are some who claim that the curanto is more evidence to suggest the theory that asserts that there was contact between America and Polynesia in the pre-Columbian epoch, given that almost all Polynesian cultures have similar cooking methods known variously as "imu", "umu", "lovo", or "hangi".

## Fish Ball

Fish balls are a common food in southern China and overseas Chinese communities made from surimi. They are also common in Scandinavia, where they are usually made from cod or haddock.

***Figure:*** *Fish ball*

### Production

Fish balls made in Scandinavia are similar to meatballs, only with fish instead of pork or beef.

Meatballs made in Asia differ significantly in texture from their European counterparts. Instead of grinding and forming meats, meat used for making meatballs is pounded, which lends a smooth texture to the meatballs. This is also often the case for fillings in steamed dishes. Pounding, unlike grinding, uncoils and stretches previously wound and tangled protein strands in meat.

### Regional Variations

***Figure:*** *Steamed rice rolls with fish balls*

***Faroe Islands:*** In Faroe Islands, fish balls are called *knettir* and are made with ground fish and fat.

***Fuzhou:*** In the Fuzhou area, "Fuzhou fish balls" (□yÞ] | œ8N) are made from fish and has minced pork filling within the fish ball.

***Hong Kong:*** There are two kinds of fishballs sold in Hong Kong: yellow and white.

***Yellow Fish Ball (Street Food):*** Smaller in size, made from cheaper fish meat than white fish balls, they are usually sold at food stalls with five to seven fish balls on a bamboo skewer. The fish balls are usually boiled in a spicy curry sauce, with virtually every street stall creates their own recipes of curry satay sauce to differentiate their fish balls from other sellers. Fish balls are one of the most popular and representative "street foods" (W^-˜Ÿqß˜) of Hong Kong.

***Figure:*** *Swedish* fiskbullar, *here served with dill sauce and pasta*

***Figure:*** Kaeng khiao wan luk chin pla*: Thai green curry with fish balls*

To reduce cost, yellow fish balls sold by street food stalls consist of less than 20% fish meat, and they are mass-produced in large quantities by factory machines to cater to the large consumptive needs of the people of Hong Kong. The fish used in these factory produced fish balls are not selected under close supervision for quality and freshness, and monosodium glutamate (MSG) is added for flavour.

Yellow fishballs are distinct from the white fish balls served in restaurants, which are more expensive to produce and have a lesser proportion of flour, in terms of texture and flavour. Additionally, to produce white fish balls by hand would require more preparation and work on the part of the chef to ensure the quality of the fish balls produced before every business day.

Many of the stall owners who sell yellow fish balls depend on them as their sole support, similar to the hot dog stands in the United States.

### History

Diet is never only used to allay one's hunger and slake one's thirst. Since factors like one's economic situation, geographical location and religion vary among countries, the diet of each has its own special characteristics. The characteristic of each country's food is reflected in the areas' unique cultural background. In Hong Kong, the first fish ball was made and sold by hawkers between the 50's to 60's. Fish balls were produced and sold in this manner as a cheap, yet tasty and filling snack for the people. Although society has drastically changed since then, fish balls are still closely related to Hong Kong and its people in terms of its ubiquity, the flavour, the method of sales and their production. Subsequently, fish balls have spread throughout other parts of Asia.

### Changes between Years

Before the 1960s, people were faced with financial difficulties so they changed their occupation into becoming a hawker. However, since 1972, Hong Kong implemented "The Hong Kong Cleanup" action (nTo™™/nK□ÕR) in order to establish the sense of belonging for the local people and to keep a clean environment. After the improvement of the public's spirit and the environment, the uncleanliness and hygienic problems caused by hawkers became a focal point. From 1979 onward, the government stopped licensing hawkers for the reason of road blocking, and requested hawkers to relinquish their licenses through compensatory means in order to lessen the number of existing hawkers. In 1995, the government established two new policies which officially prohibited the existence of hawkers and fined those with licenses. In 2000, despite the fact that the government had loosened its supervisory control of hawkers in New Territories, the recent incident caused by arresting hawkers revealed their determination in executing this policy.

The historical developments related to hawkers would directly affect the development of the fish ball. The ban against hawkers had changed the operation mode from vendors to street stalls. Nowadays, most of the people would prefer to buy them from the street stalls as opposed to the vendors.

### Ingredients of Fish Balls

The basic ingredients are fish and sometimes flour; flavourings, such as salt and sugar, can also be used. The proportion of fish and flour depends on the quality and type of fish balls to be made. The white fish balls found in some traditional Hong Kong restaurants are made using only fresh fish; while the street fried fish balls are made by using cheap fish and a mixture of flour in order to reduce costs from the wholesale business. In the past, a wide variety and good quality of fish was used. But now the supply quality has become tensed for some reasons. The commonly used fish to make fish balls include Flathead mullet (]NÍhZ›/ÏpÔNZ›) and Daggertooth pike conger (€•Tœ).

Fish ball, the most popular and most common street food, was founded between the 50's and 60's. At that time, in order to reduce costs, the process of making fish balls was most likely done by mixing and frying the remaining materials of ChouZhou fish ball noÞ]}vZ›8N or stale fishes. Nowadays, fish balls sold in Hong Kong are mainly imported by wholesale businesses. Therefore, the texture of the fish ball does not vary too widely. In recent years, fish balls are sold with different hot or curry sauces.

### Prices of Fish Ball

*Selling price:*

In general, the current market price in 2012 ranges from HKD$6 to HKD$9 per stick (with 5 fish balls); from HKD$15 to HKD$20 per small bowl (with 10 fish balls); and around HKD$30 per big bowl (with about 20 fish balls). Based on the current situation, we calculate that the price per fish ball is about HKD$1.5, which has increased by 50% compared to 5 years ago. For some stalls, the price has increased even more. For instance, in Tai Wai, the price jumped from $5 to $8 or an increase of 60% within two years.

### Wholesale Price

When you consider the wholesale price of $16/kg (about 80 fish balls), the cost per fish ball is $0.2. By selling one fish ball, a stall owner can earn $1.3. According to the owner of Jinwei %msT, a famous fish ball stall operating its business in Mong Kok, Tsim Sha Tsui and Causeway Bay, he stated that the number of fish balls sold per day is 15,000. So the estimated net profit per day is almost $20,000, which is quite a pleasing amount.

### Unprevailing Price

Some stalls in Hong Kong are selling fish balls at an unexpected high price to tourists. For each bowl of fish balls sold to customers, the number of fish balls can vary. This just depends on the type of customer, i.e. local people or tourists. According to a reporter of SUN Life, he visited one of these stalls in Mong Kok and reported that an Australian tourist paid $40 for a big bowl containing 25 fish balls; while a local person paid $20 for a small bowl but had 30 fish balls. Obviously, the stall owner was unethical and cheated the tourist. This stall only sells 'a bowl of fish balls' and no 'fish balls on a stick'. In this way, customers will have a difficult time discovering the truth that there is no standardized number of fish balls to be sold in each bowl. Indeed, the stall workers just randomly put fish balls in the bowl thus the number contained in each bowl varies from time to time. Even for two separate local customers, they were still given different numbers of fish balls for the same price.

### *Rent*

The main reason behind the increased pricing of fish balls is because of the continuous and crazy rise of rental cost. Particularly, the rent is extremely high in some tourist-prone areas such as Causeway Bay, Mong Kok and Tsim Sha Tsui. One example is the fish ball stall located in Yee Woo Street, Causeway Bay. The rent in 2010 increased to $300,000 per month. Although there was a large flow of paying customers at this stall, the revenue earned was not enough to cover the rental cost. In the end, the owner had no choice and was forced to shut down the business.

In some special places, such as flower market, the rental cost is even higher. According to the reporter of Sharp Daily, the weekly rent of a fish ball stall increased from $380,000 in 2011 to $510,000 in 2012, which was a 34% increase. Based on the market situation in 2012, the net profit for each fish ball sold was approximately $1.1, that means the stall owner needed to sell at least 70,000 fish balls per day in order to break-even. Considering the cost of labour, ingredients and other expenses, the stall owner estimated that he has to sell a stick of fish balls every 5 seconds to break-even.

### *Target Customers*

The targeted group of customers include lower class, working class and middle class. The price level is still acceptable and affordable by these groups of people. Most of the working-class people are accustomed to eating fish balls after work or after dinner. They care less about manners and hygienic issues. They don't mind eating on the street nor concerned about the seemingly unhygienic street food. On the other hand, due to the low quality of fish ball ingredients, perceived bad eating manners and unsafe hygienic operations and the eating environment, the high income classed people or celebrities seldom or never buy from the stalls on the street. They need to protect their social status and avoid breaking the shared norm or values created among the collective groups in Hong Kong, as described in the sociological imagination. Once they break this social norm, they will appear in the news or in a magazine within a few days. For instance, Fiona Sit, Ken Hung and Vincy Chan were once being photographed by a paparazzi while they were eating fish balls on the street.

### *Innovative Way in Engaging Fish Ball Business*

To cope with the challenge of rising rental costs in Hong Kong, fish ball stall owners try to use different strategies in an attempt to increase their profit margin. In 2004, Ng Han Wai, successfully innovated a fish ball vending machine and in the same year started his own business. The first fish ball vending machine was located in the Sha Tin MTR station. Customers can buy a cup of fish balls by inserting a $5 coin or using the Octopus card which is quite convenient. All the processes are machine-automated to ensure a good quality of hygiene. From this innovative way of selling fish balls, we can see that Ng had changed the traditional way of doing business, a rationalized and humanized mode (fish ball stalls) to a new rationalized and mechanical mode (fish ball vending machine) of doing business. Under this new system, everything is done electronically with standardized selling and processing procedures. Human error can be greatly avoided.

The rent for these machines are definitely lower than the street stall's. According to Ng, each machine sold about 80 cups of fish ball per day, together with the revenue from advertisement and franchisee fees, he estimated that it only took half and a year to break-even.

Unfortunately, he tried to commit suicide in 2005 due to some personal reasons, causing all his business come to a standstill.

### White Fish Ball

White fish balls are larger in size and made with only fish, no other ingredients are added, and then boiled till done. As a result of this cooking method, these fish balls are white in colour. A good fish ball should have an elastic (bouncy) and fluffy texture and a strong taste of fish. They are made using a more costly fish, and has a considerably different texture and taste. This kind is usually eaten as a compliment with noodles at Chiuchow-style noodle restaurants, and at some *cha chaan tengs*, which also sellbeef balls ([r8N) and cuttlefish balls (¨XZ›8N). Readily available in traditional markets and supermarkets, fish balls are also a popular ingredient for hot pot.

Traditional Hong Kong fish ball restaurants (€W[_†Z›Ë†—^) all agree that "good fish" is the key in making good fish balls. They insist on using certain kinds of fresh fish and not adding flour. ]NÍh fish are said to be good for giving it a strong taste of fish (Z›sT). However, these fish are now becoming rare and hard to catch. Other fish, such as }kˆ fish, are becoming more expensive. These traditional restaurants are all facing the same problem, limited, specific fish supplies and increasing ingredient costs. The cost of these fish supplies have increased two to threefold in recent years. White fish balls from the traditional fish ball restaurants are regarded as the true fish ball because they have a higher quality. They are only made from fresh fish and only fresh fish. These fish balls are normally hand-made (KbSb) by the owners using traditional techniques. The price of these fish balls is usually higher. However, using the same ingredients, hand-made fish balls are still far better than the machine-made ones.

### Indonesia

In Indonesia, fish balls are called *bakso ikan* (fish *bakso*). The most popular bakso are made of beef, but fish bakso is also available, served with tofu and fish *otak-otak* in clear broth soup as *tahu kok*, or thinly sliced as additional ingredients in *mie goreng*, *kwetiau goreng*, and *cap cai*. A similar dish made of fish is called *pempek*.

### Peninsular Malaysia and Singapore

Fish balls are cooked in many ways in Peninsular Malaysia and Singapore. They can be served with soup and noodles like the Chiuchow style or with *yong tau foo*. There is also a type called fish ball *mee pok*.

### Philippines

The most commonly eaten type of fish balls is colloquially known simply as fish balls. It is somewhat flat in shape and most often made from the meat of cuttlefish or pollock and served with a sweet and spicy sauce or with a thick, black, sweet and sour sauce.

Fish balls in the Philippines are sold by street vendors pushing wooden deep-frying carts. The balls are served skewered, offered with a choice of three kinds of dipping sauces: spicy (white/orange coloured) - vinegar, water, diced onions and garlic, sweet (brown gravy coloured) - corn starch, banana ketchup, sugar and salt, and sweet/sour (amber or deeper orange coloured) - the sweet variety with lots of small hot chilis added. Dark sauces are rare, as these are soy sauce-based and soy sauce is expensive in terms of food cost for street food. A recent trend in the Philippine fish ball industry is the introduction of 'ball' varieties: chicken, squid (cuttlefish actually), and *kikiam*. The last are low cost renditions vaguely resembling the original Chinese delicacy of the same (soundwise) name.

### Scandinavia

*Fiskbullar* in Sweden and *fiskeboller* in Norway are usually bought in cans. In Sweden, they are normally served with mashed potatoes or rice, boiled green peas and dill, caviar or seafood sauces. In Norway, they are commonly served with potatoes and white sauce made with the stock from the can, sometimes with added curry.

### Thailand

In Thai cuisine, fish balls are also very popular. They are usually fried or grilled to be eaten as a snack. In Chinese-influenced restaurants, fish balls are cooked in noodle soups and come in many varieties. They can also be eaten in a Thai curry. *Kaeng khiao wan luk chin pla* is green curry with fish balls.

### Fish Ball Store and Social Media

Fish ball stores appeared in social media with multiple images. They do not only represent the unique eating culture of Hong Kong but also act as famous attractions in Hong Kong.

### Famous Attractions

***Big Golden Fish Ball - Cheung Chau:*** This is a very famous snack in Cheung Chau. It is a must-have item for tourists visiting Cheung Chau. Many visitors have shared their comments on Cheung Chau's fish balls in online forums, such as Open Rice.

### *Selling Price*

Due to the inflation in Hong Kong, the price has risen more than 60% within 2 years. In 2010, two Big Golden fish balls were sold at $6. In 2012, they were selling for $10.

### *Unique Selling Points*

There are several unique selling points of big golden fish balls compared to the normal ones sold on the street.

***Large in Size:*** As reflected from its name, the size is larger; they can be fist-sized.

***Special Sourcing:*** Big Golden fish ball is served with a special curry sauce.

***Texture is Al Dente:*** Big Golden fish ball is mainly made from fresh fish which makes the texture more al dente than the cheap fish ball usually sold.

Examples - famous fish ball stores

1. Kam Wing Tai Fish Ball Store G/F, 106 San Hing Street, Cheung Chau
2. Sun Jiu Kee Snacks Store 3 Tung Wan Street, Cheung Chau
3. Cheung Chau Cheung Kee 83A Praya Street, Cheung Chau

## Fish Slice

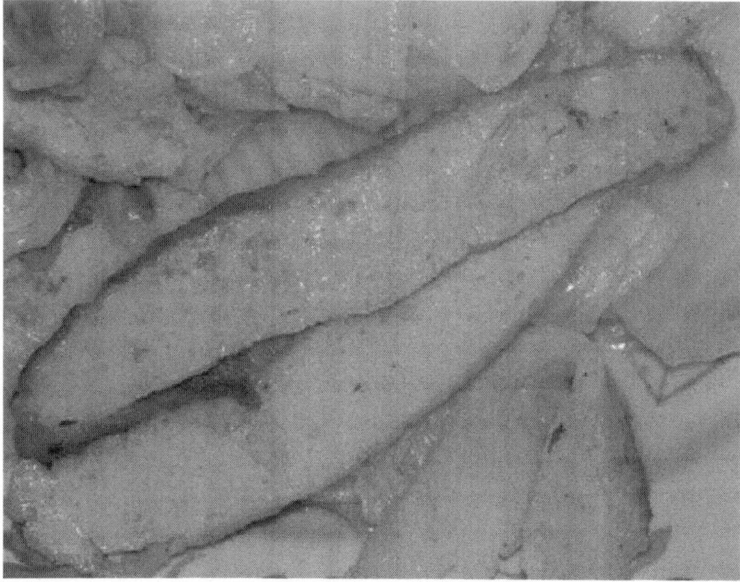

***Figure:*** *Fish slice*

Fish cake or fish slice is a commonly cooked food in southern China and overseas Chinese communities. The fillet is made of fish that has been finely pulverized. It is made of the same surimi used to make fish balls.

### *Usage*

When production is finished, the fish usually comes in a large frozen rectangular fillet. Prior to cooking, the large rectangular fish mold must be defrosted and cut into smaller slices.

### *Hong Kong*

Fish slices are often used in Chinese cuisine, especially in noodle dishes. They are commonly found in dai pai dong. Perhaps the most popular use of fish slices is mixing with fish balls or wonton noodles.

Another usage in home cooking is to combine them with soft vegetables like cabbage or bean sprouts. Meats like pork are also sometimes used for plate-dishes.

## Fish Chowder

Fish chowder is a type of chowder made with salmon, cod, or other fish. It is similar to the more popular clam chowder. Common ingredients include potatoes, onions, celery, corn, carrots, and bacon.

### *Gefilte Fish*

***Figure:*** *Gefilte fish: whole stuffed & garnished fish*

Gefilte fish is an Ashkenazi Jewish dish made from a poached mixture of ground boned fish, such as carp, whitefish or pike, which is typically eaten as an appetizer.

Although the dish historically consisted of a minced-fish forcemeat stuffed inside the fish skin, as its name implies, since the 19th century the skin has commonly been omitted and the seasoned fish is formed into patties similar to *quenelles* or fish balls. They are popular on Shabbat and Holidays such as Passover, although they may be consumed throughout the year.

### Preparation and Serving

Traditionally, carp, pike, mullet, or whitefish were used to make *gefilte fish*, but more recently other fish with white flesh such as Nile Perch have been used, and there is a pink variation using salmon. There are even vegetarian variations.

Ingredients require selecting a fish (preferably a fresh water fish such as carp or whitefish) that is preferably at least 3 kilograms (6.6 lb) in weight. Also required are 1 kilogram (2.2 lb) of brown cooking onions, salt, pepper, and 3 to 5 eggs. Up to 200 millilitres (6.8 US fl oz) of vegetable oil (traditionally sunflower oil) may also be added if the fish is lean.

The fish is deboned and the flesh mixed with ingredients, including bread crumbs or matza meal, and fried onion. Cooking takes as much as 3 hours in traditional recipes, although in modern recipes the cooking time is often briefer.

The resultant log-shaped mixture is sliced, and usually served cold or at room temperature. Often, each slice is topped with a slice of carrot, with a horseradish mixture called *khreyn* on the side.

Due to the general poverty of the Jewish population in Eastern Europe, the 'economic' recipe for the above also may have included extra ground and soaked matza meal or bread crumbs creating many more "spare" fish balls. This form of preparation eliminated the need for picking out fish bones at the table, and "stretched" the fish further, so that even poor, but often large, families could enjoy fish on *Shabbat*. Not only is picking bones religiously prohibited on the Sabbath, but many of the commonly used fish such as carp are exceptionally bony and difficult to eat easily in whole form.

### Variations

*Gefilte fish* may be slightly sweet or savory. Preparation of *gefilte fish* with sugar or black pepper is considered an indicator of whether a Jewish community was Galitzianer (with sugar) or Litvak (with pepper), hence the boundary separating northern from southern East Yiddish has been dubbed "the Gefilte Fish Line".

## *Ready-to-Serve*

***Figure:*** *Gefilte fish: fish balls*

***Figure:*** *Jars of Gefilte fish*

The post-WW2 method of making gefilte fish commercially takes the form of patties or balls, or utilises a wax paper casing around a "log" of ground fish, which is then poached or baked. This product is sold in cans and glass jars, and packed in jelly made from fish broth. Sodium is a relatively high 220–290 mg/serving. Low-salt, low-carb, low-cholesterol, sugar-free, and kosher varieties are available. The U.S. Patent #3,108,882 "Method for Preparing an Edible Fish Product" for this jelly, which allowed mass-market distribution of gefilte fish, was granted on October 29, 1963 to Monroe Nash. Gefilte fish are also sold frozen in 'logs'.

## Symbolism

Among religiously observant Jews, *gefilte fish* has become a traditional Shabbat food to avoid *borer*, which is one of the 39 activities prohibited on Shabbat outlined in the Shulchan Aruch. Borer, literally "selection/choosing," would occur when one picks the bones out of the fish, taking "the chaff from within the food."

A less common belief is that fish are not subject to "ayin hara" ("evil eye") because they are submerged while alive, so that a dish prepared from several fish varieties brings good luck.

Fish is parve, neither milk nor meat, and according to kosher law, it may be eaten at both meat and dairy meals, although according to halakha fish and meat should not be eaten together.

## Sushi

Sushi is a Japanese food consisting of cooked vinegared rice *sushi-meshi* (¨›ï˜, "sushi rice") combined with other ingredients (*neta* [ÿ[øSÍ0¿0]), usually raw fish or other seafood. Neta and forms of sushi presentation vary widely, but the ingredient which all sushi have in common is vinegared rice. The rice is also referred to as (*shari* [W0ƒ0Š0]) and "sumeshi" (b'ï˜, "vinegared rice").

Raw meat (usually but not necessarily seafood) sliced and served by itself is *sashimi*. Many non-Japanese use the terms sashimi and sushi interchangeably, but the two dishes are actually distinct and separate. Sushi refers to any dish made with vinegared rice.

The original type of sushi, known today as *nare-zushi* (´™Œ0ÿ[øS, Ÿqÿ[øS,) was first made in Southeast Asia, possibly along what is now known as the Mekong River. The sushi cuisine then spread to Southern China before it was introduced in Japan.

The term *sushi* comes from an archaic grammatical form no longer used in other contexts; literally, *sushi* means "sour-tasting", a reflection of its historic fermented roots. The oldest form of sushi in Japan, *narezushi*, still very closely resembles this process, wherein fish is fermented via being wrapped in soured fermenting rice. The fish proteins break down via fermentation into its constituent amino acids. The fermenting rice and fish results in a sour taste and also one of the five basic tastes, called *umami* in Japanese. Contemporary Japanese sushi has little resemblance to the traditional lacto-fermented rice dish. Originally, when the fermented fish was taken out of the rice, only the fish was consumed while the fermented rice was discarded. The strong-tasting and smelling *funazushi*, a kind of *narezushi* made near Lake Biwa in Japan, resembles the traditional fermented dish. Beginning in the Muromachi period (AD 1336–1573) of Japan, vinegar was added to the mixture for better taste and preservation. The vinegar accentuated the rice's sourness and was known to increase its shelf life, allowing the fermentation process to be shortened and eventually abandoned. In the following centuries, sushi in Osaka evolved into *oshi-zushi*. The seafood and rice were pressed using wooden (usually bamboo) molds. By the mid 18th century, this form of sushi had reached Edo (contemporary Tokyo).

The contemporary version, internationally known as "sushi", was created by Hanaya Yohei (1799–1858) at the end of the Edo period in Edo. Sushi invented by Hanaya was an early form of fast food that was not fermented (therefore prepared quickly) and could be conveniently eaten with one's hands. Originally, this sushi was known as *Edomae zushi* because it used freshly caught fish in the *Edo-mae* (Edo Bay or Tokyo Bay). Though the fish used in modern sushi no longer usually comes from Tokyo Bay, it is still formally known as *Edomae nigirizushi*.

The *Oxford English Dictionary* notes the earliest written mention of sushi in English in an 1893 book, *A Japanese Interior*, where it mentions sushi as "*a roll of cold rice with fish, sea-weed, or some other flavouring*". However, there is also mention of sushi in a Japanese-English dictionary from 1873, and an 1879 article on Japanese cookery in the journal *Notes and Queries*.

### Types

The common ingredient across all kinds of sushi is vinegared *sushi rice*. Variety arises from fillings, toppings, condiments, and preparation. Traditional versus contemporary methods of assembly may create very different results from very similar ingredients. In spelling sushi, its first letter *s* is replaced with *z* when a prefix is attached, as in nigirizushi, due to consonant mutation called rendaku in Japanese.

### Chirashizushi

***Figure:*** *Chirashizushi with raw ingredients*

*Chirashizushi* ("scattered sushi") is a bowl of sushi rice topped with a variety of sashimi and garnishes (also refers to *barazushi*). *Edomae chirashizushi* (Edo-style scattered sushi) is an uncooked ingredient that is arranged artfully on top of the sushi rice in a bowl. *Gomokuzushi* (Kansai-style sushi) consists of cooked or uncooked ingredients mixed in the body of rice in a bowl. There is no set formula for the ingredients; they are either chef's choice or specified by the customer. It is commonly eaten because it is filling, fast and easy to make. *Chirashizushi* often varies regionally. It is eaten annually on Hinamatsuri in March.

### Inarizushi

*Inarizushi* is a pouch of fried tofu typically filled with sushi rice alone. It is named after the Shinto god *Inari*, who is believed to have a fondness for fried tofu. The pouch is normally fashioned as deep-fried tofu (*abura age*). Regional variations include pouches made of a thin omelette (*fukusa-zushi*, or *chakin-zushi*). It should not be confused with *inari maki*, which is a roll filled with flavoured fried tofu.

***Figure:*** *Inarizushi*

A version of *inarizushi* that includes green beans, carrots, and gobo along with rice, wrapped in a triangular aburage (fried tofu) piece, is a Hawaiian speciality, where it is called cone sushi and is often sold in *okazu-ya* (Japanese delis) and as a component of bento boxes.

### Makizushi

***Figure:*** *Futomaki*

***Figure:*** *Makizushi in preparation*

*Makizushi* ("rolled sushi"), *Norimaki* ("Nori roll") or *Makimono* ("variety of rolls") is a cylindrical piece, formed with the help of a bamboo mat, called a *makisu*. *Makizushi* is generally wrapped in nori (seaweed), but is occasionally wrapped in a thin omelette, soy paper, cucumber, or shiso (perilla) leaves. *Makizushi* is usually cut into six or eight pieces, which constitutes a single roll order. Below are some common types of makizushi, but many other kinds exist.

*Futomaki* ("thick, large or fat rolls") is a large cylindrical piece, with nori on the outside. A typical *futomaki* is five to six centimetres (2–2.5 in) in diameter. They are often made with two, three, or more fillings that are chosen for their complementary tastes and colours. During the evening of the Setsubun festival, it is traditional in the Kansai region to eat uncut futomaki in its cylindrical form, where it is called *ehô-maki* (lit. happy direction rolls). By 2000 the custom had spread to all of Japan. Futomaki are often vegetarian, and may utilise strips of cucumber, *kampyô* gourd, *takenoko* bamboo shoots, or lotus root. Strips of *tamagoyaki* omelette, tiny fish roe, chopped tuna, and *oboro* whitefish flakes are typical non-vegetarian fillings.

*Hosomaki* ("thin rolls") is a small cylindrical piece, with the nori on the outside. A typical *hosomaki* has a diameter of about two and a half centimetres (1 in). They generally contain only one filling, often tuna, cucumber, *kanpyô*, thinly sliced carrots, or, more recently, avocado. *Kappamaki,* a kind of *Hosomaki* filled with cucumber, is named after the Japanese legendary water imp fond of cucumbers called the kappa. Traditionally, *Kappamaki* is consumed to clear the palate between eating raw fish and other kinds of food, so that the flavours of the fish are distinct from the tastes of other foods. *Tekkamaki* is a kind of *Hosomaki* filled with raw tuna. Although it is believed that the name "Tekka", meaning 'red hot iron', alludes to the colour of the tuna flesh or salmon flesh, it actually originated as a quick snack to eat in gambling dens called "Tekkaba", much like the sandwich. *Negitoromaki* is a kind of *Hosomaki* filled with scallion (negi) and chopped tuna (toro). Fatty tuna is often used in this style. *Tsunamayomaki* is a kind of *Hosomaki* filled with canned tuna tossed with mayonnaise.

*Temaki* ("hand roll") is a large cone-shaped piece of nori on the outside and the ingredients spilling out the wide end. A typical *temaki* is about ten centimetres (4 in) long, and is eaten with fingers because it is too awkward to pick it up with chopsticks. For optimal taste and texture, *Temaki* must be eaten quickly after being made because the nori cone soon absorbs moisture from the filling and loses its crispness and becomes somewhat difficult to bite. For this reason, the nori in pre-made or take-out temaki is sealed in plastic film which is removed immediately before eating.

*Uramaki,* ("inside-out roll") is a medium-sized cylindrical piece with two or more fillings. *Uramaki* differs from other *makimono* because the rice is on the outside and the nori inside. The filling is in the centre surrounded by nori, then a layer of rice, and an outer coating of some other ingredients such as *roe* or toasted sesame seeds. It can be made with different fillings, such as tuna, crab meat, avocado, mayonnaise, cucumber or carrots. In Japan, urimaki is an uncommon type of makimono because of the outer layer of rice can be quite difficult to handle with fingers.

### Narezushi

*Narezushi* ("matured sushi") is a traditional form of fermented sushi. Skinned and gutted fish are stuffed with salt, placed in a wooden barrel, doused with salt again, then weighed down with a heavy tsukemonoishi (pickling stone). As days pass, water seeps out and is removed. After six months, this *sushi* can be eaten, remaining edible for another six months or more. The most famous variety of narezushi still being produced is *funa-zushi* (made from fish of the crucian carp genus, authentically from *C. auratus grandoculis* (*nigoro-buna*) endemic to Lake Biwa), a typical dish of Shiga Prefecture.

***Figure:*** *funa-zushi* (*narezushi* made from a crucian carp species)

### Nigirizushi

***Figure:*** *Sea urchin roe "gunkanmaki"*

*Nigirizushi* ("hand-pressed sushi") consists of an oblong mound of *sushi rice* that the chef presses into a small rectangular box between the palms of the hands, usually with a bit of *wasabi*, and a topping (the *neta*) draped over it. Neta are typically fish such as salmon, tuna or other seafood. Certain toppings are typically bound to the rice with a thin strip of nori, most commonly octopus (*tako*), freshwater eel (*unagi*), sea eel (*anago*), squid (*ika*), and sweet egg (*tamago*). When ordered separately, nigiri is generally served in pairs. A sushi set (a sampler dish) may contain only one piece of each topping.

*Gunkanmaki*, ("warship roll") is a special type of *nigirizushi*: an oval, hand-formed clump of sushi rice that has a strip of "nori" wrapped around its perimeter to form a vessel that is filled with some soft, loose or fine-chopped ingredient that requires the confinement of *nori* such as roe, *nattô*, oysters, sea urchin, corn with mayonnaise, and quail eggs. *Gunkanmaki* was invented at the *Ginza Kyubey* restaurant in 1941; its invention significantly expanded the repertoire of soft toppings used in sushi.

*Temarizushi* ("ball sushi") is a ball-shaped sushi made by pressing rice and fish into a ball-shaped form by hand using a plastic wrap.

### Oshizushi

**Figure:** *Sasazushi, a type of oshizushi*

*Oshizushi* ("pressed sushi"), also known as, *hako-zushi*, "box sushi"), is a pressed sushi from the Kansai region, a favourite and speciality of Osaka. A block-shaped piece formed using a wooden mold, called an oshibako. The chef lines the bottom of the oshibako with the toppings, covers them with sushi rice, and then presses the lid of the mold down to create a compact, rectilinear block. The block is removed from the mold and then cut into bite-sized pieces. Particularly famous is *battera*, pressed mackerel sushi or *saba zushi*.

### *Western-Style Sushi*

***Figure:*** *Roll-style Western sushi*

The increasing popularity of *sushi* around the world has resulted in variations typically found in the Western world, but rarely in Japan (a notable exception to this is the use of salmon which was introduced by the Norwegians in the early 1980s).

Such creations to suit the Western palate were initially fuelled by the invention of the California roll. A wide variety of popular rolls has evolved since. Though the menu names of dishes often vary by restaurant, some examples include:

| *Sushi roll name* | *Definition* |
|---|---|
| Alaska roll | a variant of the California roll with raw salmon on the inside, or layered on the outside. |
| British Columbia roll | contains grilled or barbecued salmon skin, cucumber, sweet sauce, sometimes with roe. Also sometimes referred to as salmon skin rolls outside of British Columbia, Canada. |

*Contd...*

| ***Sushi roll name*** | ***Definition*** |
|---|---|
| California roll | consists of avocado, kani kama (imitation crab/crab stick) (also can contain real crab in 'premium' varieties), cucumber and tobiko, often made *uramaki* (with rice on the outside, nori on the inside) |
| Dynamite roll | includes yellowtail (hamachi) and/or prawn tempura, and fillings such as bean sprouts, carrots, avocado, cucumber, chili, spicy mayonnaise, and roe. |
| Hawaiian roll | contains shoyu tuna (canned), tamago, kanpyô, kamaboko, and the distinctive red and green hana ebi (shrimp powder). |
| Philadelphia roll | consists of raw or smoked salmon, cream cheese (often Philadelphia cream cheese brand), cucumber or avocado, and/or onion. |
| Rainbow roll | is a California roll with typically 6–7 types of sashimi (yellowtail, tuna, salmon, snapper, white fish, eel, etc.) and avocado wrapped around it. |
| Seattle roll | consists of cucumber, avocado, and raw or smoked salmon. |
| Mango roll | includes fillings such as avocado, crab meat, tempura shrimp, mango slices, and topped off with a creamy mango paste. |
| Spider roll | includes fried soft-shell crab and other fillings such as cucumber, avocado, daikon sprouts or lettuce, roe, and sometimes spicy mayonnaise. |
| Michigan roll | includes fillings such as spicy tuna, smelt roe, spicy sauce, avocado, and sushi rice. Is a variation on Spicy Tuna Roll. |

Other rolls may include chopped scallops, spicy tuna, beef or chicken teriyaki roll, okra, and assorted vegetables such as cucumber and avocado. Sometimes sushi rolls are made with brown rice and black rice, which appear in Japanese cuisine as well. An inside-out roll allows the consumer to drape sashimi on top of the entire roll. Examples include the rainbow roll (an inside-out topped with thinly sliced *maguro, hamachi, ebi, sake* and avocado) and the caterpillar roll (an inside out topped with thinly sliced avocado). Also commonly found is the rock and roll (an inside out roll with barbecued freshwater eel and avocado with toasted sesame seeds on the outside) and the tempura roll where shrimp tempura is in the roll or an entire roll is battered and fried tempura style. In the Southern United States, many sushi restaurants prepare rolls using crawfish. Futomaki roll is found widely within Japan and is often mistaken for California roll. Other types of Western-style sushi are also rarely seen in Japan.

### Ingredients

All sushi has a base of specially prepared rice, complemented with other ingredients.

### Sushi-Meshi

*Sushi-meshi* is a preparation of white, short-grained, Japanese rice mixed with a dressing consisting of rice vinegar, sugar, salt, and occasionally kombu and *sake*. It has to be cooled to room temperature before being used for a filling in a *sushi* or else it will get too sticky while being seasoned. Traditionally, the mixing is done with a hangiri, which is a round, flat-bottom wooden tub or barrel, and a wooden paddle (shamoji).

Sushi rice is prepared with short-grain Japanese rice, which has a consistency that differs from long-grain strains such as those from India, Pakistan, Thailand, and Vietnam. The essential quality is its stickiness or glutinousness. Freshly harvested rice (shinmai) typically contains too much water, and requires extra time to drain the rice cooker after washing. In some fusion cuisine restaurants, short grain brown rice and wild rice are also used. There are regional variations in sushi rice and, of course, individual chefs have their individual methods. Most of the variations are in the rice vinegar dressing: the Kantô region (or East Japan) version of the dressing commonly uses more salt; in Kansai region (or West Japan), the dressing has more sugar.

### Nori

The black seaweed wrappers used in *makimono* are called nori. Nori is a type of algae, traditionally cultivated in the harbors of Japan. Originally, algae was scraped from dock pilings, rolled out into thin, edible sheets, and dried in the sun, in a process similar to making rice paper. Today, the commercial product is farmed, processed, toasted, packaged, and sold in sheets.

The size of a nori sheet influences the size of maki-mono. A full size sheet produces futomaki, and a half produces hosomaki and temaki. To produce gunkan and some other makimono, an appropriately sized piece of nori is cut from a whole sheet.

***Figure:*** *Sheets of nori.*

Nori by itself is an edible snack and is available with salt or flavoured with teriyaki sauce. The flavoured variety, however, tends to be of lesser quality and is not suitable for sushi.

When making *fukusazushi*, a paper-thin omelette may replace a sheet of nori as the wrapping. The omelette is traditionally made on a rectangular omelette pan (makiyakinabe), and used to form the pouch for the rice and fillings.

### Neta

***Figure:*** *Sushi with raw meat (usually pork or beef) is a rare variation occasionally seen in Japan.*

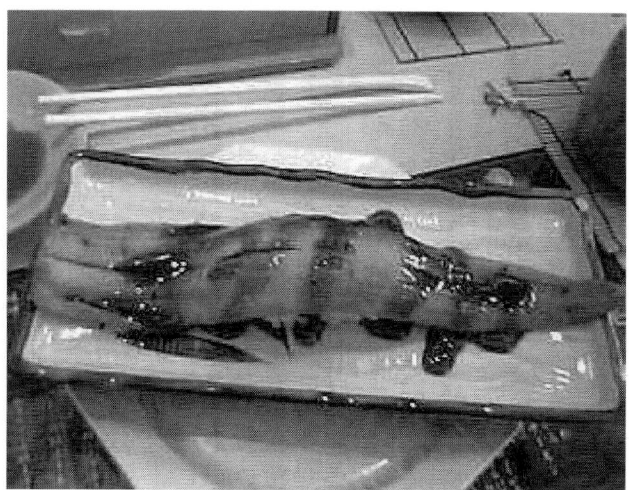

***Figure:*** *Yaki anago-ippon-nigiri. A roasted and sweet sauced whole conger.*

For culinary, sanitary, and aesthetic reasons, the minimum quality and freshness of fish to be eaten raw must be superior to that of fish which is to be cooked. Sushi chefs are trained to recognise important attributes, including smell, colour, firmness, and freedom from parasites that may go undetected in commercial inspection.

***Figure:*** *Ebifurai-Maki. Fried-shrimp roll.*

Commonly used fish are tuna (*maguro, shiro-maguro*), Japanese amberjack, yellowtail (*hamachi*), snapper (*kurodai*), mackerel (*saba*), and salmon (*sake*). The most valued sushi ingredient is *toro,* the fatty cut of the fish. This comes in a variety of *ôtoro* (often from the bluefin species of tuna) and *chûtoro,* meaning 'middle toro', implying that it is halfway into the fattiness between toro and the regular cut. *Aburi* style refers to nigiri sushi where the fish is partially grilled (topside) and partially raw. Most nigiri sushi will be completely raw.

Other seafoods such as squid (*ika*), eel (*anago* and *unagi*), pike conger (*hamo*), octopus (*tako*), shrimp (*ebi* and *amaebi*), clam (*mirugai, aoyagi* and *akagai*), fish roe (*ikura, masago, kazunoko* and *tobiko*), sea urchin (*uni*), crab (*kani*), and various kinds of shellfish (abalone, prawn, scallop) are the most popular seafoods in sushi. Oysters, however, are less common, as the taste is not thought to go well with the rice. *Kani kama*, or imitation crab stick, is commonly substituted for real crab, most notably in California rolls.

Pickled daikon radish (*takuan*) in shinko maki, pickled vegetables (*tsukemono*), fermented soybeans (*nattô*) in nattô maki, avocado, cucumber in kappa maki, asparagus, yam, pickled ume (*umeboshi*), gourd (*kanpyô*), burdock (*gobo*), and sweet corn (possibly mixed with mayonnaise) are also used in sushi.

Tofu and eggs (in the form of slightly sweet, layered omelette called *tamagoyaki* and raw quail eggs ride as a gunkan-maki topping) are common.

***Figure:*** *Date-Maki. Futomaki wrapped with sweet-tamagoyaki.*

### Condiments

Sushi is commonly eaten with condiments. Sushi may be dipped in shôyu, soy sauce, and may be flavoured with wasabi, a piquant paste made from the grated root of the *Wasabia japonica* plant. Japanese-style mayonnaise is a common condiment in Japan on salmon, pork and other sushi cuts.

True wasabi has anti-microbial properties and may reduce the risk of food poisoning. The traditional grating tool for wasabi is a sharkskin grater or *samegawa oroshi.* An imitation wasabi (*seiyo-wasabi*), made from horseradish, mustard powder and green dye is common. It is found at lower-end kaiten zushi restaurants, in bento box sushi and at most restaurants outside Japan. If manufactured in Japan, it may be labelled "Japanese Horseradish".

Gari (sweet, pickled ginger) is eaten with sushi to both cleanse the palate and aid in digestion. In Japan, green tea (*ocha*) is invariably served together with sushi. Better sushi restaurants often use a distinctive premium tea known as *mecha.* In sushi vocabulary, green tea is known as *agari.* Sushi may be garnished with Gobo, grated daikon radish, thinly sliced vegetables, carrots/radishes/cucumbers that have been shaped to look like flowers, real flowers and/or seaweed salad.

When closely arranged on a tray, different pieces are often separated by green strips called *baran* or *kiri-zasa.* Originally these were cut leaves from, respectively, the Aspidistra elatior and Sasa veitchii plants, but today the strips are usually made from green plastic.

### Nutrition

***Figure:*** *Sushi in shops are usually sold in plastic trays.*

The main ingredients of traditional Japanese sushi, raw fish and rice, are naturally low in fat, high in protein, carbohydrates, vitamins, and minerals, as are gari and nori. Other vegetables wrapped within the sushi also offer various vitamins and minerals. Many of the seafood ingredients also contain omega-3 fatty acids, which have a variety of health benefits.

### Health Risks

Some of the ingredients in sushi can present health risks. Large marine apex predators such as tuna (especially bluefin) can harbor high levels of methylmercury, which can lead to mercury poisoning when consumed in large quantity or when consumed by certain higher-risk groups including women who might get pregnant, pregnant women, nursing mothers and young children.

Sashimi or other types of sushi containing raw fish present a risk of infection by three main types of parasites:

- *Clonorchis sinensis* a fluke which can cause clonorchiasis
- *Anisakis*, a roundworm which can cause anisakiasis
- *Diphyllobothrium*, a tapeworm which can cause diphyllobothriasis

For the above reasons, the EU regulations forbids the use of fresh raw fish. It must be frozen at temperatures below "20 °C ("4 °F) in all parts of the product for no less than 24 hours. As such, a number of fishing boats, suppliers and end users "super freeze" fish for sushi to temperatures as low at "60°C. As well as parasite destruction, super freezing also prevents oxidation of the blood in tuna flesh thus preventing the discolouration that happens at temperatures above "20°C.

Some forms of sushi, notably those containing pufferfish fugu and some kinds of shellfish, can cause severe poisoning if not prepared properly. Particularly, fugu consumption can be fatal - a few deaths occur in Japan every year from eating fugu fish sushi. Fugu fish has a lethal dose of tetrodotoxin in its internal organs and must be prepared by a licensed fugu chef who has passed the prefectural examination in Japan. The licensing examination process consists of a written test, a fish-identification test, and a practical test, preparing and eating the fish. Only about 35 percent of the applicants pass. The Emperor of Japan is forbidden to eat fugu, as it is considered too risky.

### Presentation

Traditionally, sushi is served on minimalist Japanese-style, geometric, mono- or duo-tone wood or lacquer plates, in keeping with the aesthetic qualities of this cuisine. Many sushi restaurants offer fixed-price sets, selected by the chef from the catch of the day. These are often graded as *shô-chiku-bai*, shô/*matsu* (pine), chiku/*take* (bamboo) and bai/*ume*, with *matsu* the most expensive and *ume* the cheapest.

Sushi may be served *kaiten zushi* (sushi train) style. Colour-coded plates of sushi are placed on a conveyor belt; as the belt passes customers choose as they please. After finishing, the bill is tallied by counting how many plates of each colour have been taken. Newer kaiten zushi restaurants use barcodes or RFID tags embedded in the dishes to manage elapsed time after the item was prepared.

### Glossary

Some specialised or slang terms are used in the sushi culture. Most of these terms are used only in sushi bars.

- Agari: "Rise up" Green tea. Ocha (J06*f*) in usual Japanese.
- Gari: Sweet, pickled and sliced ginger, or sushi ginger. Shoga in standard Japanese.
- Gyoku: "Jewel", Sweet and cubic-shaped omelette. Tamagoyaki (uS<q, ‰sP[<q in standard Japanese.)
- Murasaki: Soy sauce. Murasaki is the colour name for violet or purple. Shoyu in standard Japanese.
- Neta: Toppings on nigiri or fillings in makimono. Ne-ta is from reversal of ta-ne. Tane (.z) in standard Japanese.
- Oaiso: "Compliment", Bill or check. Oaiso may be used in not only sushi bars but also izakaya. Okanjo or chekku in standard Japanese.
- Otemoto: Chopsticks. Otemoto means the nearest thing from the customer seated. Hashi (¸{) or ohashi in standard Japanese.
- Sabi: Japanese horseradish. Contracted form of wasabi.
- Shari: Vinegar rice or rice. It may originally be from Sanskrit (zaali 6>2?) meaning rice and/or Úarîra. Gohan (T0ï˜) or meshi (ï˜) in standard Japanese.
- Tsume: Sweet thick sauce mainly made of soy sauce. Nitsume in standard Japanese.

### Etiquette

Unlike sashimi, which is almost always eaten with chopsticks, nigirizushi is traditionally eaten with the fingers, even in formal settings. While it is commonly served on a small platter with a side dish for dipping, sushi can also be served in a *bento*, a box with small compartments that hold the various dishes of the meal.

Soy sauce is the usual condiment, and sushi is normally served with a small sauce dish, or a compartment in the bento. Traditional etiquette suggests that the sushi is turned over so that only the topping is dipped; this is because the soy sauce is for flavouring the topping, not the rice, and because the rice would absorb too much soy sauce and would fall apart. If it is difficult to turn the sushi upside-down, one can baste the sushi in soy sauce using *gari* (sliced ginger) as a brush. Toppings which have their own sauce (such as eel) should not be eaten with soy sauce.

Traditionally, the sushi chef will add an appropriate amount of wasabi to the sushi while preparing it, and etiquette suggests eating the sushi as is, since the chef is supposed to know the proper amount of wasabi to use. However, today wasabi is more a matter of personal taste, and even restaurants in Japan may serve wasabi on the side for customers to use at their discretion, even when there is wasabi already in the roll.

***Preparation Utensils***

| ***Utensil*** | ***Definition*** |
|---|---|
| Fukin | Kitchen cloth |
| Hangiri | Rice barrel |
| Hocho | Kitchen knives |
| Makisu | Bamboo rolling mat |
| Ryoribashi | Cooking chopsticks |
| Shamoji | Wooden rice paddle |
| Makiyakinabe | Rectangular omelette pan |
| Oshizushihako | a mold used to make oshizushi |

## Tuna Fish Sandwich

A tuna fish sandwich, or tuna sandwich, is a type of sandwich usually made with ingredients such as mayonnaise. Tuna salad includes other ingredients such as celery. Other common variations include the tuna boat and tuna melt.

The sandwich has been called "the mainstay of almost everyone's American childhood", and "the staple of the snatched office lunch for a generation." In the United States, 52% of canned tuna is used for sandwiches.

### *Ingredients and Preparation*

A tuna fish sandwich is usually made with canned tuna mixed with mayonnaise and other ingredients, such as chopped celery, hard-boiled eggs, cucumber, sweetcorn, pickles, onion, and/or black olives. Other recipes may use olive oil, salad cream, mustard, or yogurt instead of mayonnaise.

## Variations

***Figure:*** *A tuna melt sandwich served with French fries*

***Figure:*** *An open faced tuna sandwich with guacamole and cherry tomatoes*

A tuna melt has melted cheese on top of the tuna or on a tomato slice and is served on toasted bread. A tuna boat is a tuna sandwich served in a hot dog bun.

### Nutritional Information

Tuna is a relatively high protein food and it is very high in omega-3 fatty acids. A sandwich made from 100 grams of tuna and two slices of toasted white bread has approximately 287 Calories, 96 Calories of which are from fat. It also has 20 grams of protein and 27 grams of carbohydrates.

***According to the StarKist Company:*** The nutritional content of albacore may vary naturally from catch to catch. In particular, the fat and calorie content will differ depending on the region or depth where the fish are caught. Albacore tuna that swim close to the water surface can be higher in fat than tuna caught in deep ocean waters—but they are also lower in mercury content and higher in omega-3 content. Because availability of albacore tuna may vary from season to season, we use two different labels to accurately show the fat and calorie content of the product contents.

A larger, commercially-prepared tuna fish sandwich has more calories than noted above, based on its serving size. A 6-inch Subway tuna sub of 238 grams has 480 calories, 210 of those from fat, 600 miligrams of sodium, and 20 grams of protein.

# 2

# Fish Biology

A fish is any member of a paraphyletic group of organisms that consist of all gill-bearing aquatic craniate animals that lack limbs with digits. Included in this definition are the living hagfish, lampreys, and cartilaginous and bony fish, as well as various extinct related groups. Most fish are ectothermic ("cold-blooded"), allowing their body temperatures to vary as ambient temperatures change, though some of the large active swimmers like white shark and tuna can hold a higher core temperature.

Fish are abundant in most bodies of water. They can be found in nearly all aquatic environments, from high mountain streams (e.g., char and gudgeon) to the abyssal and even hadal depths of the deepest oceans (e.g., gulpers and anglerfish). At 32,000 species, fish exhibit greater species diversity than any other group of vertebrates.

Fish are an important resource for humans worldwide, especially as food. Commercial and subsistence fishers hunt fish in wild fisheries or farm them in ponds or in cages in the ocean. They are also caught by recreational fishers, kept as pets, raised by fishkeepers, and exhibited in public aquaria. Fish have had a role in culture through the ages, serving as deities, religious symbols, and as the subjects of art, books and movies.

Because the term "fish" is defined negatively, and excludes the tetrapods (i.e., the amphibians, reptiles, birds and mammals) which descend from within the same ancestry, it is paraphyletic, and is not considered a proper grouping in systematic biology. The traditional term pisces (also ichthyes) is considered a typological, but not a phylogenetic classification.

The earliest organisms that can be classified as fish were soft-bodied chordates that first appeared during the Cambrian period. Although they lacked a true spine, they possessed notochords which allowed them to be more agile than their invertebrate counterparts. Fish would continue to evolve through the Paleozoic era, diversifying into a wide variety of forms.

Many fish of the Paleozoic developed external armor that protected them from predators. The first fish with jaws appeared in the Silurian period, after which many (such as sharks) became formidable marine predators rather than just the prey of arthropods.

## Evolution

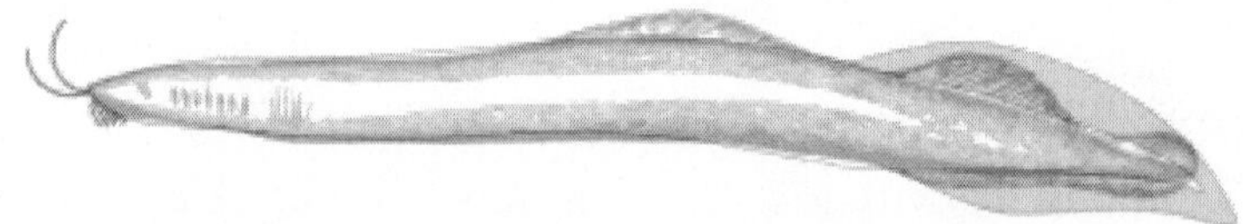

***Figure:*** *Fish do not represent a monophyletic group, and therefore the "evolution of fish" is not studied as a single event.*

Early fish from the fossil record are represented by a group of small, jawless, armored fish known as Ostracoderms. Jawless fish lineages are mostly extinct. An extant clade, the Lampreys may approximate ancient pre-jawed fish. The first jaws are found in Placodermi fossils. The diversity of jawed vertebrates may indicate the evolutionary advantage of a jawed mouth. It is unclear if the advantage of a hinged jaw is greater biting force, improved respiration, or a combination of factors.

Fish may have evolved from a creature similar to a coral-like Sea squirt, whose larvae resemble primitive fish in important ways. The first ancestors of fish may have kept the larval form into adulthood (as some sea squirts do today), although perhaps the reverse is the case.

### *Taxonomy*

Fish are a paraphyletic group: that is, any clade containing all fish also contains the tetrapods, which are not fish. For this reason, groups such as the "Class Pisces" seen in older reference works are no longer used in formal classifications.

Traditional classification divide fish into three extant classes, and with extinct forms sometimes classified within the tree, sometimes as their own classes:

- Class Agnatha (jawless fish)
  - o Subclass Cyclostomata (hagfish and lampreys)
  - o Subclass Ostracodermi (armoured jawless fish) †
- Class Chondrichthyes (cartilaginous fish)
  - o Subclass Elasmobranchii (sharks and rays)
  - o Subclass Holocephali (chimaeras and extinct relatives)
- Class Placodermi (armoured fish) †
- Class Acanthodii ("spiny sharks", sometimes classified under bony fishes)†
- Class Osteichthyes (bony fish)
  - o Subclass Actinopterygii (ray finned fishes)
  - o Subclass Sarcopterygii (fleshy finned fishes, ancestors of tetrapods)

The above scheme is the one most commonly encountered in non-specialist and general works. Many of the above groups are paraphyletic, in that they have given rise to successive groups: Agnathans are ancestral to Chondrichthyes, who again have given rise to Acanthodiians, the ancestors of Osteichthyes. With the arrival of phylogenetic nomenclature, the fishes has been split up into a more detailed scheme, with the following major groups:

- Class Myxini (hagfish)
- Class Pteraspidomorphi † (early jawless fish)
- Class Thelodonti †
- Class Anaspida †
- Class Petromyzontida or Hyperoartia
  - o Petromyzontidae (lampreys)
- Class Conodonta (conodonts) †
- Class Cephalaspidomorphi † (early jawless fish)
  - o (unranked) Galeaspida †
  - o (unranked) Pituriaspida †
  - o (unranked) Osteostraci †
- Infraphylum Gnathostomata (jawed vertebrates)
  - o Class Placodermi † (armoured fish)
  - o Class Chondrichthyes (cartilaginous fish)
  - o Class Acanthodii † (spiny sharks)

- o Superclass Osteichthyes (bony fish)
  - – Class Actinopterygii (ray-finned fish)
  - – Subclass Chondrostei
  - – Order Acipenseriformes (sturgeons and paddlefishes)
  - – Order Polypteriformes (reedfishes and bichirs).
  - – Subclass Neopterygii
  - – Infraclass Holostei (gars and bowfins)
  - – Infraclass Teleostei (many orders of common fish)
  - – Class Sarcopterygii (lobe-finned fish)
  - – Subclass Actinistia (coelacanths)
  - – Subclass Dipnoi (lungfish)

† – indicates extinct taxon

Some palaeontologists contend that because Conodonta are chordates, they are primitive fish.

The position of hagfish in the phylum chordata is not settled. Phylogenetic research in 1998 and 1999 supported the idea that the hagfish and the lampreys form a natural group, the Cyclostomata, that is a sister group of the Gnathostomata.

The various fish groups account for more than half of vertebrate species. There are almost 28,000 known extant species, of which almost 27,000 are bony fish, with 970 sharks, rays, and chimeras and about 108 hagfish and lampreys. A third of these species fall within the nine largest families; from largest to smallest, these families are Cyprinidae, Gobiidae, Cichlidae, Characidae, Loricariidae, Balitoridae, Serranidae, Labridae, and Scorpaenidae. About 64 families are monotypic, containing only one species. The final total of extant species may grow to exceed 32,500.

### Diversity

The term "fish" most precisely describes any non-tetrapod craniate (i.e. an animal with a skull and in most cases a backbone) that has gills throughout life and whose limbs, if any, are in the shape of fins. Unlike groupings such as birds or mammals, fish are not a single clade but a paraphyletic collection of taxa, including hagfishes, lampreys, sharks and rays, ray-finned fish, coelacanths, and lungfish. Indeed, lungfish and coelacanths are closer relatives of tetrapods (such as mammals, birds, amphibians, etc.) than of other fish such as ray-finned fish or sharks, so the last common ancestor of all fish is also an ancestor to tetrapods. As paraphyletic groups are no longer recognised in modern systematic biology, the use of the term "fish" as a biological group must be avoided.

Many types of aquatic animals commonly referred to as "fish" are not fish in the sense given above; examples include shellfish, cuttlefish, starfish, crayfish and jellyfish. In earlier times, even biologists did not make a distinction – sixteenth century natural historians classified also seals, whales, amphibians, crocodiles, even hippopotamuses, as well as a host of aquatic invertebrates, as fish. However, according the definition above, all mammals, including cetaceans like whales and dolphins, are not fish. In some contexts, especially in aquaculture, the true fish are referred to as finfish (or fin fish) to distinguish them from these other animals.

A typical fish is ectothermic, has a streamlined body for rapid swimming, extracts oxygen from water using gills or uses an accessory breathing organ to breathe atmospheric oxygen, has two sets of paired fins, usually one or two (rarely three) dorsal fins, an anal fin, and a tail fin, has jaws, has skin that is usually covered with scales, and lays eggs.

Each criterion has exceptions. Tuna, swordfish, and some species of sharks show some warm-blooded adaptations—they can heat their bodies significantly above ambient water temperature. Streamlining and swimming performance varies from fish such as tuna, salmon, and jacks that can cover 10–20 body-lengths per second to species such as eels and rays that swim no more than 0.5 body-lengths per second. Many groups of freshwater fish extract oxygen from the air as well as from the water using a variety of different structures.

Lungfish have paired lungs similar to those of tetrapods, gouramis have a structure called the labyrinth organ that performs a similar function, while many catfish, such as *Corydoras* extract oxygen via the intestine or stomach. Body shape and the arrangement of the fins is highly variable, covering such seemingly un-fishlike forms as seahorses, pufferfish, anglerfish, and gulpers. Similarly, the surface of the skin may be naked (as in moray eels), or covered with scales of a variety of different types usually defined as placoid (typical of sharks and rays), cosmoid (fossil lungfish and coelacanths), ganoid (various fossil fish but also living gars and bichirs), cycloid, and ctenoid (these last two are found on most bony fish).

There are even fish that live mostly on land. Mudskippers feed and interact with one another on mudflats and go underwater to hide in their burrows. The catfish *Phreatobius cisternarum* lives in underground, phreatic habitats, and a relative lives in waterlogged leaf litter.

Fish range in size from the huge 16-metre (52 ft) whale shark to the tiny 8-millimetre (0.3 in) stout infantfish.

Fish species diversity is roughly divided equally between marine (oceanic) and freshwater ecosystems. Coral reefs in the Indo-Pacific constitute the centre of diversity for marine fishes, whereas continental freshwater fishes are most diverse in large river basins of tropical rainforests, especially the Amazon, Congo, and Mekong basins. More than 5,600 fish species inhabit Neotropical freshwaters alone, such that Neotropical fishes represent about 10% of all vertebrate species on the Earth. Exceptionally rich sites in the Amazon basin, such as Cantão State Park, can contain more freshwater fish species than occur in all of Europe.

## Anatomy

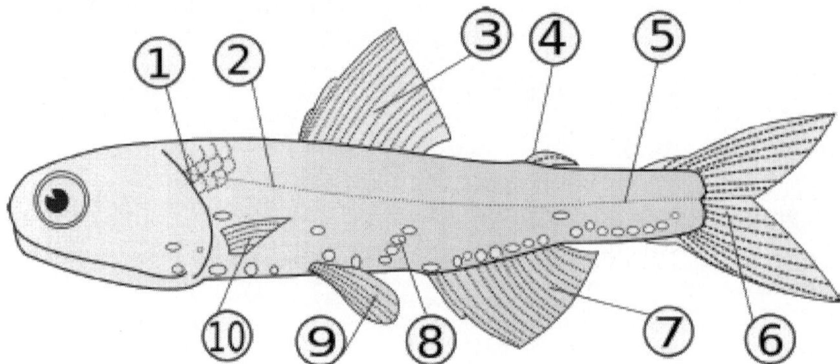


***Figure:*** *The anatomy of* Lampanyctodes hectoris

*(1) – operculum (gill cover), (2) – lateral line, (3) – dorsal fin, (4) – fat fin, (5) – caudal peduncle, (6) – caudal fin, (7) – anal fin, (8) – photophores, (9) – pelvic fins (paired), (10) – pectoral fins (paired)*

## Respiration

Most fish exchange gases using gills on either side of the pharynx. Gills consist of threadlike structures called filaments. Each filament contains a capillary network that provides a large surface area for exchanging oxygen and carbon dioxide. Fish exchange gases by pulling oxygen-rich water through their mouths and pumping it over their gills. In some fish, capillary blood flows in the opposite direction to the water, causing countercurrent exchange. The gills push the oxygen-poor water out through openings in the sides of the pharynx. Some fish, like sharks and lampreys, possess multiple gill openings. However, bony fish have a single gill opening on each side. This opening is hidden beneath a protective bony cover called an operculum.

Juvenile bichirs have external gills, a very primitive feature that they share with larval amphibians.

Fish from multiple groups can live out of the water for extended time periods. Amphibious fish such as the mudskipper can live and move about on land for up to several days, or live in stagnant or otherwise oxygen depleted water. Many such fish can breathe air via a variety of mechanisms. The skin of anguillid eels may absorb oxygen directly. The buccal cavity of the electric eel may breathe air. Catfish of the families Loricariidae, Callichthyidae, and Scoloplacidae absorb air through their digestive tracts. Lungfish, with the exception of the Australian lungfish, and bichirs have paired lungs similar to those of tetrapods and must surface to gulp fresh air through the mouth and pass spent air out through the gills. Gar and bowfin have a vascularized swim bladder that functions in the same way. Loaches, trahiras, and many catfish breathe by passing air through the gut. Mudskippers breathe by absorbing oxygen across the skin (similar to frogs). A number of fish have evolved so-called accessory breathing organs that extract oxygen from the air. Labyrinth fish (such as gouramis and bettas) have a labyrinth organ above the gills that performs this function. A few other fish have structures resembling labyrinth organs in form and function, most notably snakeheads, pikeheads, and the Clariidae catfish family.

Breathing air is primarily of use to fish that inhabit shallow, seasonally variable waters where the water's oxygen concentration may seasonally decline. Fish dependent solely on dissolved oxygen, such as perch and cichlids, quickly suffocate, while air-breathers survive for much longer, in some cases in water that is little more than wet mud. At the most extreme, some air-breathing fish are able to survive in damp burrows for weeks without water, entering a state of aestivation (summertime hibernation) until water returns.

Air breathing fish can be divided into obligate air breathers and facultative air breathers. Obligate air breathers, such as the African lungfish, *must* breathe air periodically or they suffocate. Facultative air breathers, such as the catfish *Hypostomus plecostomus*, only breathe air if they need to and will otherwise rely on their gills for oxygen. Most air breathing fish are facultative air breathers that avoid the energetic cost of rising to the surface and the fitness cost of exposure to surface predators.

### Circulation

Fish have a closed-loop circulatory system. The heart pumps the blood in a single loop throughout the body. In most fish, the heart consists of four parts, including two chambers and an entrance and exit.

The first part is the sinus venosus, a thin-walled sac that collects blood from the fish's veins before allowing it to flow to the second part, the atrium, which is a large muscular chamber.

The atrium serves as a one-way antechamber, sends blood to the third part, ventricle.

The ventricle is another thick-walled, muscular chamber and it pumps the blood, first to the fourth part, bulbus arteriosus, a large tube, and then out of the heart.

The bulbus arteriosus connects to the aorta, through which blood flows to the gills for oxygenation.

### Digestion

Jaws allow fish to eat a wide variety of food, including plants and other organisms. Fish ingest food through the mouth and break it down in the esophagus. In the stomach, food is further digested and, in many fish, processed in finger-shaped pouches called pyloric caeca, which secrete digestive enzymes and absorb nutrients.

Organs such as the liver and pancreas add enzymes and various chemicals as the food moves through the digestive tract. The intestine completes the process of digestion and nutrient absorption.

### Excretion

As with many aquatic animals, most fish release their nitrogenous wastes as ammonia. Some of the wastes diffuse through the gills. Blood wastes are filtered by the kidneys.

Saltwater fish tend to lose water because of osmosis. Their kidneys return water to the body.

The reverse happens in freshwater fish: they tend to gain water osmotically. Their kidneys produce dilute urine for excretion. Some fish have specially adapted kidneys that vary in function, allowing them to move from freshwater to saltwater.

### Scales

The scales of fish originate from the mesoderm (skin); they may be similar in structure to teeth.

## Sensory and Nervous System

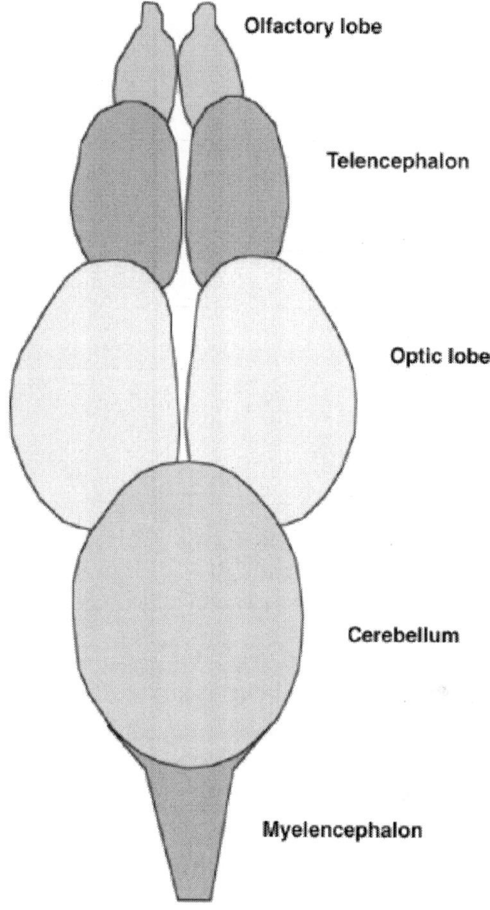


***Figure:*** *Dorsal view of the brain of the rainbow trout*

### Central Nervous System

Fish typically have quite small brains relative to body size compared with other vertebrates, typically one-fifteenth the brain mass of a similarly sized bird or mammal. However, some fish have relatively large brains, most notably mormyrids and sharks, which have brains about as massive relative to body weight as birds and marsupials.

Fish brains are divided into several regions. At the front are the olfactory lobes, a pair of structures that receive and process signals from the nostrils via the two olfactory nerves. The olfactory lobes are very large in fish that hunt primarily by smell, such as hagfish, sharks, and catfish. Behind the olfactory lobes is the two-lobed telencephalon, the structural equivalent to the cerebrum in higher vertebrates. In fish the telencephalon is concerned mostly with olfaction. Together these structures form the forebrain.

Connecting the forebrain to the midbrain is the diencephalon (in the diagram, this structure is below the optic lobes and consequently not visible). The diencephalon performs functions associated with hormones and homeostasis. The pineal body lies just above the diencephalon. This structure detects light, maintains circadian rhythms, and controls colour changes.

The midbrain or mesencephalon contains the two optic lobes. These are very large in species that hunt by sight, such as rainbow trout and cichlids. The hindbrain or metencephalon is particularly involved in swimming and balance. The cerebellum is a single-lobed structure that is typically the biggest part of the brain. Hagfish and lampreys have relatively small cerebellae, while the mormyrid cerebellum is massive and apparently involved in their electrical sense.

The brain stem or myelencephalon is the brain's posterior. As well as controlling some muscles and body organs, in bony fish at least, the brain stem governs respiration and osmoregulation.

### Sense Organs

Most fish possess highly developed sense organs. Nearly all daylight fish have colour vision that is at least as good as a human's. Many fish also have chemoreceptors that are responsible for extraordinary senses of taste and smell. Although they have ears, many fish may not hear very well.

Most fish have sensitive receptors that form the lateral line system, which detects gentle currents and vibrations, and senses the motion of nearby fish and prey. Some fish, such as catfish and sharks, have organs that detect weak electric currents on the order of millivolt. Other fish, like the South American electric fishes Gymnotiformes, can produce weak electric currents, which they use in navigation and social communication.

Fish orient themselves using landmarks and may use mental maps based on multiple landmarks or symbols. Fish behaviour in mazes reveals that they possess spatial memory and visual discrimination.

### Vision

Vision is an important sensory system for most species of fish. Fish eyes are similar to those of terrestrial vertebrates like birds and mammals, but have a more spherical lens. Their retinas generally have both rod cells and cone cells (for scotopic and photopic vision), and most species have colour vision. Some fish can see ultraviolet and some can see polarized light. Amongst jawless fish, the lamprey has well-developed eyes, while the hagfish has only primitive eyespots. Fish vision shows adaptation to their visual environment, for example deep sea fishes have eyes suited to the dark environment.

### *Hearing*

Hearing is an important sensory system for most species of fish. Fish sense sound using their lateral lines and their ears.

### *Capacity for Pain*

Experiments done by William Tavolga provide evidence that fish have pain and fear responses. For instance, in Tavolga's experiments, toadfish grunted when electrically shocked and over time they came to grunt at the mere sight of an electrode.

In 2003, Scottish scientists at the University of Edinburgh and the Roslin Institute concluded that rainbow trout exhibit behaviours often associated with pain in other animals. Bee venom and acetic acid injected into the lips resulted in fish rocking their bodies and rubbing their lips along the sides and floors of their tanks, which the researchers concluded were attempts to relieve pain, similar to what mammals would do. Neurons fired in a pattern resembling human neuronal patterns.

Professor James D. Rose of the University of Wyoming claimed the study was flawed since it did not provide proof that fish possess "conscious awareness, particularly a kind of awareness that is meaningfully like ours". Rose argues that since fish brains are so different from human brains, fish are probably not conscious in the manner humans are, so that reactions similar to human reactions to pain instead have other causes. Rose had published a study a year earlier arguing that fish cannot feel pain because their brains lack a neocortex. However, animal behaviourist Temple Grandin argues that fish could still have consciousness without a neocortex because "different species can use different brain structures and systems to handle the same functions."

Animal welfare advocates raise concerns about the possible suffering of fish caused by angling. Some countries, such as Germany have banned specific types of fishing, and the British RSPCA now formally prosecutes individuals who are cruel to fish.

### *Muscular System*

Most fish move by alternately contracting paired sets of muscles on either side of the backbone. These contractions form S-shaped curves that move down the body. As each curve reaches the back fin, backward force is applied to the water, and in conjunction with the fins, moves the fish forward. The fish's fins function like an airplane's flaps. Fins also increase the tail's surface area, increasing speed. The streamlined body of the fish decreases the amount of friction from the water. Since body tissue is denser than water, fish must compensate for the difference or they will sink. Many bony fish have an internal organ called a swim bladder that adjusts their buoyancy through manipulation of gases.

### *Homeothermy*

Although most fish are exclusively ectothermic, there are exceptions.

Certain species of fish maintain elevated body temperatures. Endothermic teleosts (bony fish) are all in the suborder Scombroidei and include the billfishes, tunas, and one species of "primitive" mackerel (*Gasterochisma melampus*). All sharks in the family Lamnidae – shortfin mako, long fin mako, white, porbeagle, and salmon shark – are endothermic, and evidence suggests the trait exists in family Alopiidae (thresher sharks). The degree of endothermy varies from the billfish, which warm only their eyes and brain, to bluefin tuna and porbeagle sharks who maintain body temperatures elevated in excess of 20 °C above ambient water temperatures. Endothermy, though metabolically costly, is thought to provide advantages such as increased muscle strength, higher rates of central nervous system processing, and higher rates of digestion.

### *Reproductive System*

Fish reproductive organs include testes and ovaries. In most species, gonads are paired organs of similar size, which can be partially or totally fused. There may also be a range of secondary organs that increase reproductive fitness.

In terms of spermatogonia distribution, the structure of teleosts testes has two types: in the most common, spermatogonia occur all along the seminiferous tubules, while in Atherinomorph fish they are confined to the distal portion of these structures. Fish can present cystic or semi-cystic spermatogenesis in relation to the release phase of germ cells in cysts to the seminiferous tubules lumen.

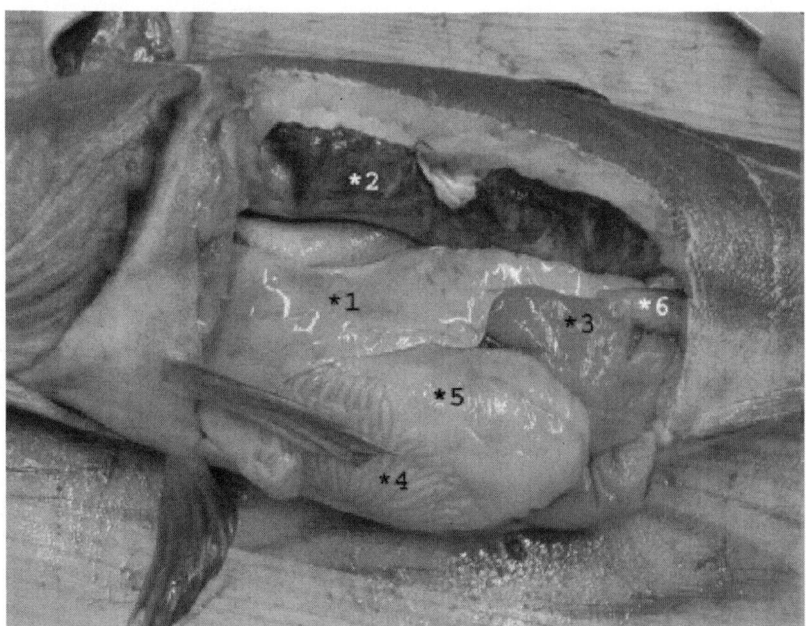


***Figure:*** *Organs: 1. Liver, 2. Gas bladder, 3. Roe, 4. Pyloric caeca, 5. Stomach, 6. Intestine*

Fish ovaries may be of three types: gymnovarian, secondary gymnovarian or cystovarian. In the first type, the oocytes are released directly into the coelomic cavity and then enter the ostium, then through the oviduct and are eliminated. Secondary gymnovarian ovaries shed ova into the coelom from which they go directly into the oviduct. In the third type, the oocytes are conveyed to the exterior through the oviduct. Gymnovaries are the primitive condition found in lungfish, sturgeon, and bowfin. Cystovaries characterize most teleosts, where the ovary lumen has continuity with the oviduct. Secondary gymnovaries are found in salmonids and a few other teleosts.

Oogonia development in teleosts fish varies according to the group, and the determination of oogenesis dynamics allows the understanding of maturation and fertilization processes. Changes in the nucleus, ooplasm, and the surrounding layers characterize the oocyte maturation process.

Postovulatory follicles are structures formed after oocyte release; they do not have endocrine function, present a wide irregular lumen, and are rapidly reabsorbed in a process involving the apoptosis of follicular cells. A degenerative process called follicular atresia reabsorbs vitellogenic oocytes not spawned. This process can also occur, but less frequently, in oocytes in other development stages.

Some fish are hermaphrodites, having both testes and ovaries either at different phases in their life cycle or, as in hamlets, have them simultaneously.

Over 97% of all known fish are oviparous, that is, the eggs develop outside the mother's body. Examples of oviparous fish include salmon, goldfish, cichlids, tuna, and eels. In the majority of these species, fertilisation takes place outside the mother's body, with the male and female fish shedding their gametes into the surrounding water. However, a few oviparous fish practice internal fertilization, with the male using some sort of intromittent organ to deliver sperm into the genital opening of the female, most notably the oviparous sharks, such as the horn shark, and oviparous rays, such as skates. In these cases, the male is equipped with a pair of modified pelvic fins known as claspers.

Marine fish can produce high numbers of eggs which are often released into the open water column. The eggs have an average diameter of 1 millimetre (0.039 in).

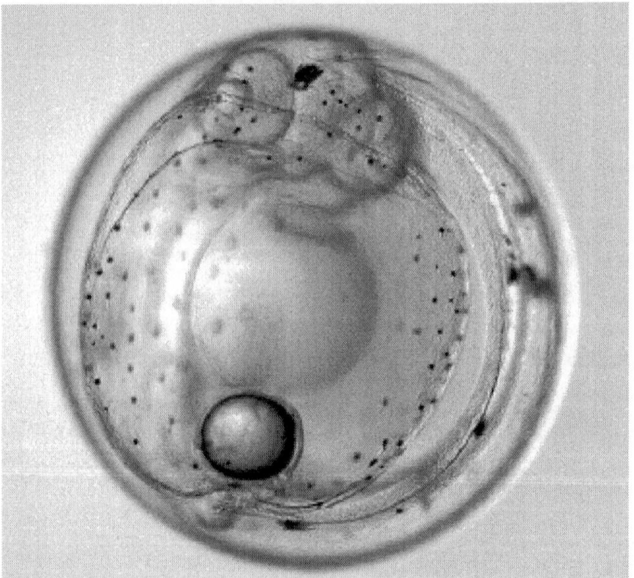

***Figure:*** *An example of zooplankton*

The newly hatched young of oviparous fish are called larvae. They are usually poorly formed, carry a large yolk sac (for nourishment) and are very different in appearance from juvenile and adult specimens. The larval period in oviparous fish is relatively short (usually only several weeks), and larvae rapidly grow and change appearance and structure (a process termed metamorphosis) to become juveniles.

During this transition larvae must switch from their yolk sac to feeding on zooplankton prey, a process which depends on typically inadequate zooplankton density, starving many larvae.

In ovoviviparous fish the eggs develop inside the mother's body after internal fertilization but receive little or no nourishment directly from the mother, depending instead on the yolk. Each embryo develops in its own egg. Familiar examples of ovoviviparous fish include guppies, angel sharks, and coelacanths.

Some species of fish are viviparous. In such species the mother retains the eggs and nourishes the embryos. Typically, viviparous fish have a structure analogous to the placenta seen in mammals connecting the mother's blood supply with that of the embryo. Examples of viviparous fish include the surf-perches, splitfins, and lemon shark. Some viviparous fish exhibit oophagy, in which the developing embryos eat other eggs produced by the mother. This has been observed primarily among sharks, such as the shortfin mako and porbeagle, but is known for a few bony fish as well, such as the halfbeak *Nomorhamphus ebrardtii*. Intrauterine cannibalism is an even more unusual mode of vivipary, in which the largest embryos eat weaker and smaller siblings. This behaviour is also most commonly found among sharks, such as the grey nurse shark, but has also been reported for *Nomorhamphus ebrardtii*.

Aquarists commonly refer to ovoviviparous and viviparous fish as livebearers.

### Diseases

Like other animals, fish suffer from diseases and parasites. To prevent disease they have a variety of defences. *Non-specific* defences include the skin and scales, as well as the mucus layer secreted by the epidermis that traps and inhibits the growth of microorganisms. If pathogens breach these defences, fish can develop an inflammatory response that increases blood flow to the infected region and delivers white blood cells that attempt to destroy pathogens. Specific defences respond to particular pathogens recognised by the fish's body, i.e., an immune response. In recent years, vaccines have become widely used in aquaculture and also with ornamental fish, for example furunculosis vaccines in farmed salmon and koi herpes virus in koi.

Some species use cleaner fish to remove external parasites. The best known of these are the Bluestreak cleaner wrasses of the genus *Labroides* found on coral reefs in the Indian and Pacific Oceans. These small fish maintain so-called "cleaning stations" where other fish congregate and perform specific movements to attract the attention of the cleaners. Cleaning behaviours have been observed in a number of fish groups, including an interesting case between two cichlids of the same genus, *Etroplus maculatus*, the cleaner, and the much larger *Etroplus suratensis*.

### Immune System

Immune organs vary by type of fish. In the jawless fish (lampreys and hagfish), true lymphoid organs are absent. These fish rely on regions of lymphoid tissue within other organs to produce immune cells. For example, erythrocytes, macrophages and plasma cells are produced in the anterior kidney (or pronephros) and some areas of the gut (where granulocytes mature.) They resemble primitive bone marrow in hagfish. Cartilaginous fish (sharks and rays) have a more advanced immune system. They have three specialised organs that are unique to chondrichthyes; the epigonal organs (lymphoid tissue similar to mammalian bone) that surround the gonads, the Leydig's organ within the walls of their esophagus, and a spiral valve in their intestine. These organs house typical immune cells (granulocytes, lymphocytes and plasma cells). They also possess an identifiable thymus and a well-developed spleen (their most important immune organ) where various lymphocytes, plasma cells and macrophages develop and are stored. Chondrostean fish (sturgeons, paddlefish and bichirs) possess a major site for the production of granulocytes within a mass that is associated with the meninges (membranes surrounding the central nervous system.) Their heart is frequently covered with tissue that contains lymphocytes, reticular cells and a small number of macrophages. The chondrostean kidney is an important hemopoietic organ; where erythrocytes, granulocytes, lymphocytes and macrophages develop.

Like chondrostean fish, the major immune tissues of bony fish (or teleostei) include the kidney (especially the anterior kidney), which houses many different immune cells. In addition, teleost fish possess a thymus, spleen and scattered immune areas within mucosal tissues (e.g. in the skin, gills, gut and gonads). Much like the mammalian immune system, teleost erythrocytes, neutrophils and granulocytes are believed to reside in the spleen whereas lymphocytes are the major cell type found in the thymus. In 2006, a lymphatic system similar to that in mammals was described in one species of teleost fish, the zebrafish. Although not confirmed as yet, this system presumably will be where naive (unstimulated) T cells accumulate while waiting to encounter an antigen.

### Conservation

The 2006 IUCN Red List names 1,173 fish species that are threatened with extinction. Included are species such as Atlantic cod, Devil's Hole pupfish, coelacanths, and great white sharks. Because fish live underwater they are more difficult to study than terrestrial animals and plants, and information about fish populations is often lacking. However, freshwater fish seem particularly threatened because they often live in relatively small water bodies. For example, the Devil's Hole pupfish occupies only a single 3 by 6 metres (10 by 20 ft) pool.

### Overfishing

Overfishing is a major threat to edible fish such as cod and tuna. Overfishing eventually causes population (known as stock) collapse because the survivors cannot produce enough young to replace those removed. Such commercial extinction does not mean that the species is extinct, merely that it can no longer sustain a fishery.

One well-studied example of fishery collapse is the Pacific sardine *Sadinops sagax caerulues* fishery off the California coast. From a 1937 peak of 790,000 long tons (800,000 t) the catch steadily declined to only 24,000 long tons (24,000 t) in 1968, after which the fishery was no longer economically viable.

The main tension between fisheries science and the fishing industry is that the two groups have different views on the resiliency of fisheries to intensive fishing. In places such as Scotland, Newfoundland, and Alaska the fishing industry is a major employer, so governments are predisposed to support it. On the other hand, scientists and conservationists push for stringent protection, warning that many stocks could be wiped out within fifty years.

### Habitat Destruction

A key stress on both freshwater and marine ecosystems is habitat degradation including water pollution, the building of dams, removal of water for use by humans, and the introduction of exotic species. An example of a fish that has become endangered because of habitat change is the pallid sturgeon, a North American freshwater fish that lives in rivers damaged by human activity.

## Environmental Impact of Fishing

The environmental impact of fishing can be divided into issues that involve the availability of fish to be caught, such as overfishing, sustainable fisheries, and fisheries management; and issues that involve the impact of fishing on other elements of the environment, such as by-catch. These conservation issues are part of marine conservation, and are addressed in fisheries science programs. There is a growing gap between how many fish are available to be caught and humanity's desire to catch them, a problem that gets worse as the world population grows. Similar to other environmental issues, there can be conflict between the fishermen who depend on fishing for their livelihoods and fishery scientists who realise that if future fish populations are to be sustainable then some fisheries must reduce or even close.

The journal *Science* published a four-year study in November 2006, which predicted that, at prevailing trends, the world would run out of wild-caught seafood in 2048. The scientists stated that the decline was a result of overfishing, pollution and other environmental factors that were reducing the population of fisheries at the same time as their ecosystems were being degraded. Yet again the analysis has met criticism as being fundamentally flawed, and many fishery management officials, industry representatives and scientists challenge the findings, although the debate continues. Many countries, such as Tonga, the United States, Australia and New Zealand, and international management bodies have taken steps to appropriately manage marine resources.

### *Overfishing*

Overfishing has also been widely reported due to increases in the volume of fishing hauls to feed a quickly growing number of consumers. This has led to the breakdown of some sea ecosystems and several fishing industries whose catch has been greatly diminished. The extinction of many species has also been reported. According to a Food and Agriculture Organisation estimate, over 70% of the world's fish species are either fully exploited or depleted. According to the Secretary General of the 2002 World Summit on Sustainable Development, "Overfishing cannot continue, the depletion of fisheries poses a major threat to the food supply of millions of people."

The cover story of the May 15, 2003 issue of the science journal *Nature* – with Dr. Ransom A. Myers, an internationally prominent fisheries biologist (Dalhousie University, Halifax, Canada) as the lead author – was devoted to a summary of the scientific information. The story asserted that, as compared with 1950 levels, only a remnant (in some instances, as little as 10%) of all large ocean-fish stocks are left in the seas. These large ocean fish are the species at the top of the food chains (e.g., tuna, cod, among others). This article was subsequently criticized as being fundamentally flawed, although much debate still exists (Walters 2003; Hampton et al. 2005; Maunder et al. 2006; Polacheck 2006; Sibert et al. 2006) and the majority of fisheries scientists now consider the results irrelevant with respect to large pelagics (the open seas).

### Ecological Disruption

Fishing may disrupt food webs by targeting specific, in-demand species. There might be too much fishing of prey species such as sardines and anchovies, thus reducing the food supply for the predators. It may also cause the increase of prey species when the target fishes are predator species such as salmon and tuna. Fisheries can reduce fish stocks that cetaceans rely on for food.

### By-Catch

By-catch is the portion of the catch that is not the target species. These are either kept to be sold or discarded. In some instances the discarded portion is known as discards.

### Possible Remedies

Many governments and intergovernmental bodies have implemented fisheries management policies designed to curb the environmental impact of fishing. Fishing conservation aims to control the human activities that may completely decrease a fish stock or washout an entire aquatic environment. These laws include the quotas on the total catch of particular species in a fishery, effort quotas (e.g., number of days at sea), the limits on the number of vessels allowed in specific areas, and the imposition of seasonal restrictions on fishing..

In 2008 a large scale study of fisheries that used individual transferable quotas and ones that didn't provided strong evidence that individual transferable quotas can help to prevent collapses and restore fisheries that appear to be in decline.

Fish farming has been proposed as a more sustainable alternative to traditional capture of wild fish. However, fish farming has been found to have negative impacts on nearby wild fish. Further, farming of predatory fish like salmon can rely on fish feed that is based on fish meal and oil from wild fish.

The environmental impact of recreational fishing may be alleviated to some extent by catch and release fishing.

## Exotic Species

Introduction of non-native species has occurred in many habitats. One of the best studied examples is the introduction of Nile perch into Lake Victoria in the 1960s. Nile perch gradually exterminated the lake's 500 endemic cichlid species.

Some of them survive now in captive breeding programmes, but others are probably extinct. Carp, snakeheads, tilapia, European perch, brown trout, rainbow trout, and sea lampreys are other examples of fish that have caused problems by being introduced into alien environments.

## Fishing Industry

The fishing industry includes any industry or activity concerned with taking, culturing, processing, preserving, storing, transporting, marketing or selling fish or fish products. It is defined by the FAO as including recreational, subsistence and commercial fishing, and the harvesting, processing, and marketing sectors.

The commercial activity is aimed at the delivery of fish and other seafood products for human consumption or as input factors in other industrial processes. Directly or indirectly, the livelihood of over 500 million people in developing countries depends on fisheries and aquaculture.

### *Sectors*

There are three principal industry sectors:

- The commercial sector: comprises enterprises and individuals associated with wild-catch or aquaculture resources and the various transformations of those resources into products for sale. It is also referred to as the "seafood industry", although non-food items such as pearls are included among its products.
- The traditional sector: comprises enterprises and individuals associated with fisheries resources from which aboriginal people derive products in accordance with their traditions.
- The recreational sector: comprises enterprises and individuals associated for the purpose of recreation, sport or sustenance with fisheries resources from which products are derived that are not for sale.

### *Commercial Sector*

The commercial sector of the fishing industry comprises the following chain:

1. Commercial fishing and fish farming which produce the fish
2. Fish processing which produce the fish products
3. Marketing of the fish products

## *World Production*

Fish are harvested by commercial fishing and aquaculture.

According to the Food and Agriculture Organisation (FAO), the world harvest in 2005 consisted of 93.3 million tonnes captured by commercial fishing in wild fisheries, plus 48.1 million tonnes produced by fish farms.

In addition, 1.3 million tons of aquatic plants (seaweed etc.) were captured in wild fisheries and 14.8 million tons were produced by aquaculture. The number of individual fish caught in the wild has been estimated at 0.97-2.7 trillion per year (not counting fish farms or marine invertebrates).

Following is a table of the 2005 world fishing industry harvest in tonnes by capture and by aquaculture.

| | *Capture* | *Aquaculture* | *Total* |
|---|---|---|---|
| Fish, crustaceans, molluscs, etc. | 93,253,346 | 48,149,792 | 141,403,138 |
| Aquatic plants | 1,305,803 | 14,789,972 | 16,095,775 |
| Total | 94,559,149 | 62,939,764 | 157,498,913 |

This equates to about 24.4 kilograms a year for the average person on Earth.

## Juvenile Fish

Fish go through various juvenile stages between birth and adulthood. They start as eggs which hatch into larva. The larva are not able to feed themselves, and carry a yolk-sac which provides their nutrition. Before the yolk-sac completely disappears, the tiny fish must become capable of feeding themselves. When they have developed to the point where are they capable of feeding themselves, the fish are called fry.

When, in addition, they have developed scales and working fins, the transition to a juvenile fish is complete and it is called a fingerling. Fingerlings are typically about the size of fingers. The juvenile stage lasts until the fish is fully grown, sexually mature and interacting with other adult fish.

***Figure:*** *This young Chinook salmon, with scales and working fins, is a* fingerling

### *Growth Stages*

Ichthyoplankton *(planktonic or drifting fish)* are the eggs and larvae of fish. They are usually found in the sunlit zone of the water column, less than 200 metres deep, sometimes called the epipelagic or photic zone. Ichthyoplankton are planktonic, meaning they cannot swim effectively under their own power, but must drift with ocean currents. Fish eggs cannot swim at all, and are unambiguously planktonic. Early stage larvae swim poorly, but later stage larvae swim better and cease to be planktonic as they grow into juveniles. Fish larvae are part of the zooplankton that eat smaller plankton, while fish eggs carry their own food supply. Both eggs and larvae are themselves eaten by larger animals.

According to Kendall et al. 1984 there are three main developmental stages of fish:

- Egg stage: Spawning to hatching. This stage is used instead of using an embryonic stage because there are aspects, such as those to do with the egg envelope, that are not just embryonic aspects.
- Larval stage: From hatching till all fin rays are present and the growth of fish scales has started (squamation). A key event

is when the notochord associated with the tail fin on the ventral side of the spinal cord develops and becomes flexible. A transitional stage, the yolk-sac larval stage, lasts from hatching to the absorption of the yolk-sac.

- Juvenile stage: Starts when the transformation or metamorphosis from larva to juvenile is complete, that is, when the larva develops the features of a juvenile fish. These features are that all the fin rays are present and that scale growth is under way. The stage completes when the juvenile becomes adult, that is, when it becomes sexually mature or starts interacting with other adults.

This article is about the juvenile stage.

- Fry – refers to a recently hatched fish that has reached the stage where its yolk-sac has almost disappeared and its swim bladder is operational to the point where the fish can actively feed for itself.
- Fingerling – refers to a fish that has reached the stage where the fins can be extended and where scales have started developing throughout the body. It this stage, the fish is typically about the size of a finger.

### Juvenile Salmon

Fry and fingerling are terms that can be applied to juvenile fish of most species. But some groups of fishes have juvenile development stages particular to the group. This section details the stages and the particular names used for juvenile salmon.

- *Sac fry* or *alevin* – The life cycle of salmon begins, and usually ends, in the backwaters of streams and rivers. These are the salmon spawning grounds, where salmon eggs are deposited, for safety, in the gravel. The salmon spawning grounds are also the salmon nurseries, providing a more protected environment than the ocean usually offers. After 2 to 6 months, the eggs hatch into tiny larvae, called *sac fry* or *alevin*. The alevin have a sac containing the remainder of the yolk, and they stay hidden in the gravel for a few days while they feed on the yolk.
- *Fry* – When the sac or yolk has almost gone, the baby fish must find food for themselves, so they leave the protection of the gravel and start feeding on plankton. At this point, the baby salmon are called *fry*.

- *Parr* – At the end of the summer, the fry develop into juvenile fish called *parr*. Parr feed on small invertebrates and are camouflaged with a pattern of spots and vertical bars. They remain in this stage for up to three years.
- *Smolt* – As they approach the time when they are ready to migrate out to the sea, the parr lose their camouflage bars and undergo a process of physiological changes, which allows them to survive a shift from freshwater to saltwater. At this point, the salmon are called *smolt*. Smolt spend time in the brackish waters of the river estuary while their body chemistry adjusts (osmoregulation) to the higher salt levels they will encounter in the ocean. Smolt also grow the silvery scales which visually confuse ocean predators.
- *Post-smolt* – When they have matured sufficiently in late spring, and are about 15 to 20 centimetres long, the smolt swim out of the rivers and into the sea. There they spend their first year as a *post-smolt*. Post-smolt form schools with other post-smolt, and set off to find deep-sea feeding grounds. They then spend up to four more years as adult ocean salmon while their full swimming and reproductive capacity develops.

***Figure:*** *Salmon enter the ocean as* post-smolt *and mature into adult salmon. They gain most of their weight in the ocean*

### Protection from Predators

Juvenile fish need protection from predators. Juvenile species, as with small species in general, can achieve some safety in numbers by schooling together. Juvenile coastal fish are drawn to turbid shallow waters and to mangrove structures, where they have better protection from predators. As the fish grow, their foraging ability increases and their vulnerability to predators decreases, and they tend to shift from mangroves to mudflats. In the open sea juvenile species often aggregate around floating objects such as jellyfish and *Sargassum* seaweed. This can significantly increase their survival rates.

### As Food

Juvenile fish are marketed as food.

- *Whitebait* is a marketing term for the fry of fish, typically between 25 and 50 millimetres long. Such juvenile fish often travel together in schools along the coast, and move into estuaries and sometimes up rivers where they can be easily caught with fine meshed fishing nets. *Whitebaiting* is the activity of catching whitebait. Whitebait are tender and edible, and can be regarded as a delicacy. The entire fish is eaten including head, fins and gut. Some species make better eating than others, and the particular species that are marketed as "whitebait" varies in different parts of the world. As whitebait consists of fry of many important food species (such as herring, sprat, sardines, mackerel, bass and many others) it is not an ecologicially viable foodstuff and in several countries strict controls on harvesting exist.
- *Elvers* are young eels. Traditionally, fishermen consumed elvers as a cheap dish, but environmental changes have reduced eel populations. Similar to whitebait, they are now considered a delicacy and are priced at up to 1000 euro per kilogram. Spain has eel dishes, called "angulas", which are baby eels usually served sauteed in olive oil, garlic and a chili pepper. Elvers are now very expensive. A small serving of angulas can cost equivalent of $US100, and there are imitation angulas which can be purchased cheaply.

## Commercial Fishing

Commercial fishing is the activity of catching fish and other seafood for commercial profit, mostly from wild fisheries. It provides a large quantity of food to many countries around the world, but those who practice it as an industry must often pursue fish far into the ocean under adverse conditions. Large-scale commercial fishing is also known as industrial fishing. This profession has gained in popularity with the development of shows such as *Deadliest Catch*, *Swords*, and *Wicked Tuna*. The major fishing industries are not only owned by major corporations but by small families as well. The industry has had to adapt through the years in order to keep earning a profit.

A study taken on some small family-owned commercial fishing companies showed that they adapted to continue to earn a living but not necessarily make a large profit. It is the adaptability of the fishermen and their methods that cause some concern for fishery managers and researchers; they say that for those reasons, the sustainability of the marine ecosystems could be in danger of being ruined.

Commercial fishermen harvest a wide variety of animals, ranging from tuna, cod, carp, and salmon to shrimp, krill, lobster, clams, squid, and crab, in various fisheries for these species.

There are large and important fisheries worldwide for various species of fish, mollusks, crustaceans, and echinoderms. However, a very small number of species support the majority of the world's fisheries. Some of these species are herring, cod, anchovy, tuna, flounder, mullet, squid, shrimp, salmon, crab, lobster, oyster and scallops. All except these last four provided a worldwide catch of well over a million tonnes in 1999, with herring and sardines together providing a catch of over 22 million metric tons in 1999. Many other species are fished in smaller numbers.

The industry, in 2006, also managed to generate over 185 billion dollars in sales and also provide over two million jobs, according to an economic report released by NOAA's Fisheries Service. Commercial fishing may offer an abundance of jobs, but the pay varies from boat to boat, season to season. Crab fisherman Cade Smith was quoted in an article by *Business Week* as saying, "There was always a top boat where the crew members raked in $50,000 during the three- to five-day king crab season—or $100,000 for the longer snow crab season". That may be true, but there are also the boats who don't do well; Smith said later in the same article that his worst season left him with a loss of 500 dollars.

A 2009 paper in *Science* estimates, for the first time, the total world fish biomass as somewhere between 0.8 and 2.0 billion tonnes.

### *Methods and Gear*

Commercial fishing uses many different methods to effectively catch a large variety of species including large nets, pole and line, trolling with single lines, and traps or pots. Sustainability of fisheries is improved by using specific equipment that eliminates or minimizes catching non-targeted species.

Fishing methods vary according to the region, the species being fished for, and the technology available to the fishermen. A commercial fishing enterprise may vary from one man with a small boat with hand-casting nets or a few pot traps, to a huge fleet of trawlers processing tons of fish every day.

Commercial fishing gears in use today include surrounding nets (e.g. purse seine), seine nets (e.g. beach seine), trawls (e.g. bottom trawl), dredges, hooks and lines (e.g. long line and handline), lift nets, gillnets, entangling nets, Pole and Line, and traps

Commercial fishing gear is specifically designed and updated to avoid catching certain species of animal that is unwanted or endangered. Billions of dollars are spent each year in researching/ developing new techniques to reduce the injury and even death of unwanted marine animals caught by the fishermen. In fact, there was a study taken in 2000 on different deterrents and how effective they are at deterring the target species. The study showed that most auditory deterrents helped prevent whales from being caught while more physical barriers helped prevent birds from getting tangled within the net.

### Occupational Risk

During 2000-2006, commercial fishing was one of the most dangerous occupations in the United States, with an average annual fatality rate of 115 deaths per 100,000 fishermen. This fatality rate is 3 times that of the next most dangerous job in the U.S. and more than 25 times that of the national average across all workers. Also, between the years of 1919 and 2005, 4111 fishermen died in fishing related accidents in the United Kingdom industry alone. These deaths are generally a result of a combination of severe weather conditions, extreme fatigue due to the fact that any one fisherman usually puts in a 21-hour shift, and dangerous equipment.

The U.S. Coast Guard has primary jurisdiction over the safety of the U.S. commercial fishing fleet, enforcing regulations of the U.S. Commercial Fishing Industry Vessel Safety Act of 1988 (CFIVSA). CFIVSA regulations focus primarily on saving lives after the loss of a vessel and not on preventing vessels from capsizing or sinking, falls overboard, or injuries on deck. CFIVSA regulations require that commercial fishing vessels carry various equipment (e.g., life rafts, radio beacons, and immersion suits) depending on the size of the vessel and the area in which it operates. Not all commercial fishermen follow safety regulations and advice. One study of Maine fishermen found that less than 25% of the fishermen interviewed had recent training in first aid or CPR, only 75% of the boats had survival suits and only 36% had a survival craft. Even the ships that did have the necessary equipment did not consistently have a captain that fully understood how to use the safety equipment.

Common causes of fishing-related deaths include vessel disasters, falls overboard, and onboard injuries. The United States National Institute of Occupational Safety and Health Commercial Fishing Incident Database found that between 2000 and 2010, most vessel disasters often were initiated by flooding, vessel instability, and large waves, and that severe weather conditions contributed to a majority of fatal vessel disasters.

Most falls overboard went unwitnessed, and in none of the cases documented was the victim wearing a personal flotation device (PFD). Onboard injuries often result when a crew member is caught in a line and pulled into a winch on deck. The installation of a readily accessible emergency stop switch on the winch can potentially prevent these kinds of injuries.

### *Fish Farming*

Aquaculture is the cultivation of aquatic organisms. Unlike fishing, aquaculture, also known as aquafarming, is the cultivation of aquatic populations under controlled conditions. Mariculture refers to aquaculture practiced in marine environments. Particular kinds of aquaculture include algaculture (the production of kelp/seaweed and other algae); fish farming; shrimp farming, shellfish farming, and the growing of cultured pearls.

Fish farming involves raising fish commercially in tanks or enclosed pools, usually for food. Fish species raised by fish farms include carp, salmon, tilapia, catfish and cod. Increasing demands on wild fisheries by commercial fishing operations have caused widespread overfishing. Fish farming offers an alternative solution to the increasing market demand for fish and fish protein.

## Mariculture

Mariculture is a specialised branch of aquaculture involving the cultivation of marine organisms for food and other products in the open ocean, an enclosed section of the ocean, or in tanks, ponds or raceways which are filled with seawater. An example of the latter is the farming of marine fish, including finfish and shellfish e.g.prawns, or oysters and seaweed in saltwater ponds. Non-food products produced by mariculture include: fish meal, nutrient agar, jewellery (e.g. cultured pearls), and cosmetics.

### Methods

***Algae:*** Seawater algae such as kelp can be farmed in at least two ways. It can be grown around a rope that is anchored to the sea floor so that they do not drift away. Off the coast of California, boats with mowers harvest the top few feet of natural kelp beds. Kelp provides alginate, an edible material used in ice cream and cosmetics. The industry also supplies the dietary supplement industry.

### Shellfish

Similar to algae cultivation, shellfish can be farmed in multiple ways: on ropes, in bags or cages, or directly on (or within) the intertidal substrate. Shellfish mariculture does not require feed or fertilizer inputs, nor insecticides or antibiotics, making shellfish aquaculture (or 'mariculture') a self-supporting system. Shellfish can also be used in multi-species cultivation techniques, where shellfish can utilise waste generated by higher trophic level organisms.

### Open Ocean

Raising marine organisms under controlled conditions in exposed, high-energy ocean environments beyond significant coastal influence, is a relatively new approach to mariculture. Open ocean aquaculture (OOA) uses cages, nets, or long-line arrays that are moored, towed or float freely. Research and commercial open ocean aquaculture facilities are in operation or under development in Australia, Chile, China, France, Ireland, Italy, Japan, Mexico, and Norway. As of 2004, two commercial open ocean facilities were operating in U.S. waters, raising Threadfin near Hawaii and cobia near Puerto Rico. An operation targeting bigeye tuna recently received final approval. All U.S. commercial facilities are currently sited in waters under state or territorial jurisdiction.

### Sea Ranching

The Japanese apply a principle based on operant conditioning and the migratory nature of certain species. The fishermen raise hatchlings in a closely knitted net in a harbor, sounding an underwater horn before each feeding. When the fish are old enough they are freed from the net to mature in the open sea. During spawning season, about 80% of these fish return to their birthplace. The fishermen sound the horn and then net those fish that respond.

### Seawater Ponds

In seawater pond mariculture, fish are raised in ponds which receive water from the sea. This has the benefit that the nutrition (e.g. microorganisms) present in the seawater can be used. This is a great advantage over traditional fish farms (e.g. sweet water farms) for which the farmers buy feed (which is expensive).

Other advantages are that water purification plants may be planted in the ponds to eliminate the buildup of nitrogen, from fecal and other contamination. Also, the ponds can be left unprotected from natural predators, providing another kind of filtering.

## Environmental Effects of Fish Farming

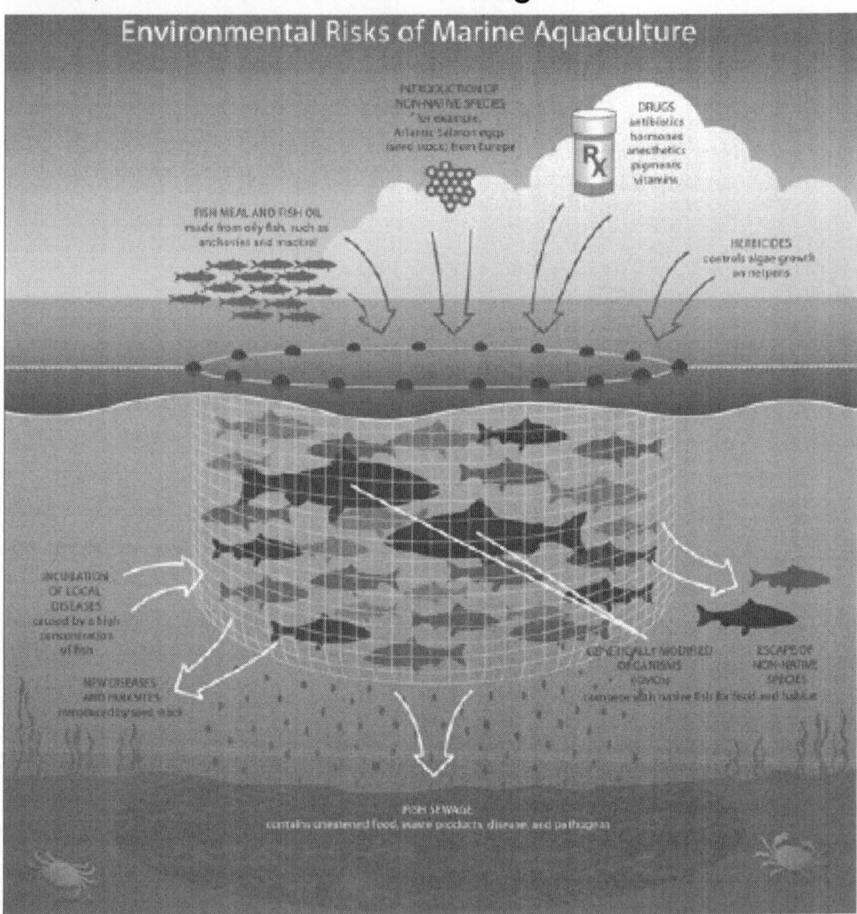


***Figure:*** *Mariculture has rapidly expanded over the last two decades due to new technology, improvements in formulated feeds, greater biological understanding of farmed species, increased water quality within closed farm systems, greater demand for seafood products, site expansion and government interest. As a consequence, mariculture has been subject to some controversy regarding its social and environmental impacts.*

Commonly identified environmental impacts from marine farms are:

1. Wastes from cage cultures;
2. Farm escapees and invasives;
3. Genetic pollution and disease and parasite transfer;
4. Habitat modification.

As with most farming practices, the degree of environmental impact depends on the size of the farm, the cultured species, stock density, type of feed, hydrography of the site, and husbandry methods. The adjacent diagram connects these causes and effects.

### Wastes from Cage Cultures

Mariculture of finfish can require a significant amount of fishmeal or other high protein food sources. Originally, a lot of fishmeal went to waste due to inefficient feeding regimes and poor digestibility of formulated feeds which resulted in poor feed conversion ratios.

In cage culture, several different methods are used for feeding farmed fish – from simple hand feeding to sophisticated computer-controlled systems with automated food dispensers coupled with *in situ* uptake sensors that detect consumption rates. In coastal fish farms, overfeeding primarily leads to increased disposition of detritus on the seafloor (potentially smothering seafloor dwelling invertebrates and altering the physical environment), while in hatcheries and land-based farms, excess food goes to waste and can potentially impact the surrounding catchment and local coastal environment. This impact is usually highly local, and depends significantly on the settling velocity of waste feed and the current velocity (which varies both spatially and temporally) and depth.

### Farm Escapees and Invasives

The impact of escapees from aquaculture operations depends on whether or not there are wild conspecifics or close relatives in the receiving environment, and whether or not the escapee is reproductively capable. Several different mitigation/prevention strategies are currently employed, from the development of infertile triploids to land-based farms which are completely isolated from any marine environment. Escapees can adversely impact local ecosystems through hybridization and loss of genetic diversity in native stocks, increase negative interactions within an ecosystem (such as predation and competition), disease transmission and habitat changes (from trophic cascades and ecosystem shifts to varying sediment regimes and thus turbidity).

The accidental introduction of invasive species is also of concern. Aquaculture is one of the main vectors for invasives following accidental releases of farmed stocks into the wild. One example is the Siberian sturgeon (*Acipenser baerii*) which accidentally escaped from a fish farm into the Gironde Estuary (Southwest France) following a severe storm in December 1999 (5,000 individual fish escaped into the estuary which had never hosted this species before). Molluscan farming is another example whereby species can be introduced to new environments by 'hitchhiking' on farmed molluscs. Also, farmed molluscs themselves can become dominate predators and/or competitors, as well as potentially spread pathogens and parasites.

### Genetic Pollution and Disease and Parasite Transfer

One of the primary concerns with mariculture is the potential for disease and parasite transfer. Farmed stocks are often selectively bred to increase disease and parasite resistance, as well as improving growth rates and quality of products. As a consequence, the genetic diversity within reared stocks decreases with every generation - meaning they can potentially reduce the genetic diversity within wild populations if they escape into those wild populations. Such genetic pollution from escaped aquaculture stock can reduce the wild population's ability to adjust to the changing natural environment. Also, maricultured species can harbour diseases and parasites (e.g., lice) which can be introduced to wild populations upon their escape. An example of this is the parasitic sea lice on wild and farmed Atlantic salmon in Canada.

Also, non-indigenous species which are farmed may have resistance to, or carry, particular diseases (which they picked up in their native habitats) which could be spread through wild populations if they escape into those wild populations. Such 'new' diseases would be devastating for those wild populations because they would have no immunity to them.

### Habitat Modification

With the exception of benthic habitats directly beneath marine farms, most mariculture causes minimal destruction to habitats. However, the destruction of mangrove forests from the farming of shrimps is of concern. Globally, shrimp farming activity is a small contributor to the destruction of mangrove forests; however, locally it can be devastating. Mangrove forests provide rich matrices which support a great deal of biodiversity – predominately juvenile fish and crustaceans. Furthermore, they act as buffering systems whereby they reduce coastal erosion, and improve water quality for in situ animals by processing material and 'filtering' sediments.

### *Others*

In addition, nitrogen and phosphorus compounds from food and waste may lead to blooms of phytoplankton, whose subsequent degradation can drastically reduce oxygen levels. If the algae are toxic, fish are killed and shellfish contaminated.

### *Sustainability*

Mariculture development must be sustained by basic and applied research and development in major fields such as nutrition, genetics, system management, product handling, and socioeconomics. One approach is closed systems that have no direct interaction with the local environment. However, investment and operational cost are currently significantly higher than open cages, limiting them to their current role as hatcheries.

### *Potential Benefits*

Sustainable mariculture promises economic and environmental benefits. Economies of scale imply that ranching can produce fish at lower cost than industrial fishing, leading to better human diets and the gradual elimination of unsustainable fisheries. Maricultured fish are also perceived to be of higher quality than fish raised in ponds or tanks, and offer more diverse choice of species. Consistent supply and quality control has enabled integration in food market channels.

## Fish Processing

The term fish processing refers to the processes associated with fish and fish products between the time fish are caught or harvested, and the time the final product is delivered to the customer. Although the term refers specifically to fish, in practice it is extended to cover any aquatic organisms harvested for commercial purposes, whether caught in wild fisheries or harvested from aquaculture or fish farming.

Larger fish processing companies often operate their own fishing fleets or farming operations. The products of the fish industry are usually sold to grocery chains or to intermediaries. Fish are highly perishable. A central concern of fish processing is to prevent fish from deteriorating, and this remains an underlying concern during other processing operations. Fish processing can be subdivided into fish handling, which is the preliminary processing of raw fish, and the manufacture of fish products. Another natural subdivision is into primary processing involved in the filleting and freezing of fresh fish for onward distribution to fresh fish retail and catering outlets, and the secondary processing that produces chilled, frozen and canned products for the retail and catering trades.

There is evidence humans have been processing fish since the early Holocene. These days, fish processing is undertaken by artisan fishermen, on board fishing or fish processing vessels, and at fish processing plants. Fish is a highly perishable food which needs proper handling and preservation if it is to have a long shelf life and retain a desirable quality and nutritional value. The central concern of fish processing is to prevent fish from deteriorating. The most obvious method for preserving the quality of fish is to keep them alive until they are ready for cooking and eating. For thousands of years, China achieved this through the aquaculture of carp. Other methods used to preserve fish and fish products include

- the control of temperature using ice, refrigeration or freezing
- the control of water activity by drying, salting, smoking or freeze-drying
- the physical control of microbial loads through microwave heating or ionizing irradiation
- the chemical control of microbial loads by adding acids
- oxygen deprivation, such as vacuum packing.

Usually more than one of these methods is used. When chilled or frozen fish or fish products are transported by road, rail, sea or air, the cold chain must be maintained. This requires insulated containers or transport vehicles and adequate refrigeration. Modern shipping containers can combine refrigeration with a controlled atmosphere.

Fish processing is also concerned with proper waste management and with adding value to fish products. There is an increasing demand for ready to eat fish products, or products that don't need much preparation.

### Handling the Catch

When fish are captured or harvested for commercial purposes, they need some preprocessing so they can be delivered to the next part of the marketing chain in a fresh and undamaged condition. This means, for example, that fish caught by a fishing vessel need handling so they can be stored safely until the boat lands the fish on shore. Typical handling processes are

- transferring the catch from the fishing gear (such as a trawl, net or fishing line) to the fishing vessel
- holding the catch before further handling
- sorting and grading
- bleeding, gutting and washing
- chilling
- storing the chilled fish
- unloading, or landing the fish when the fishing vessel returns to port

The number and order in which these operations are undertaken varies with the fish species and the type of fishing gear used to catch it, as well as how large the fishing vessel is and how long it is at sea, and the nature of the market it is supplying. Catch processing operations can be manual or automated. The equipment and procedures in modern industrial fisheries are designed to reduce the rough handling of fish, heavy manual lifting and unsuitable working positions which might result in injuries.

### *Handling Live Fish*

An alternative, and obvious way of keeping fish fresh is to keep them alive until they are delivered to the buyer or ready to be eaten. This is a common practice worldwide. Typically, the fish are placed in a container with clean water, and dead, damaged or sick fish are removed. The water temperature is then lowered and the fish are starved to reduce their metabolic rate. This decreases fouling of water with metabolic products (ammonia, nitrite and carbon dioxide) that become toxic and make it difficult for the fish to extract oxygen.

Fish can be kept alive in floating cages, wells and fish ponds. In aquaculture, holding basins are used where the water is continuously filtered and its temperature and oxygen level are controlled. In China, floating cages are constructed in rivers out of palm woven baskets, while in South America simple fish yards are built in the backwaters of rivers.

Live fish can be transported by methods which range from simple artisanal methods where fish are placed in plastic bags with an oxygenated atmosphere, to sophisticated systems which use trucks that filter and recycle the water, and add oxygen and regulate temperature.

### Preservation

Preservation techniques are needed to prevent fish spoilage and lengthen shelf life. They are designed to inhibit the activity of spoilage bacteria and the metabolic changes that result in the loss of fish quality. Spoilage bacteria are the specific bacteria that produce the unpleasant odours and flavours associated with spoiled fish. Fish normally host many bacteria that are not spoilage bacteria, and most of the bacteria present on spoiled fish played no role in the spoilage. To flourish, bacteria need the right temperature, sufficient water and oxygen, and surroundings that are not too acidic. Preservation techniques work by interrupting one or more of these needs. Preservation techniques can be classified as follows.

### Control of Temperature

***Figure:*** *Ice preserves fish and extends shelf life by lowering the temperature*

If the temperature is decreased, the metabolic activity in the fish from microbial or autolytic processes can be reduced or stopped. This is achieved by refrigeration where the temperature is dropped to about 0 °C, or freezing where the temperature is dropped below -18°C. On fishing vessels, the fish are refrigerated mechanically by circulating cold air or by packing the fish in boxes with ice. Forage fish, which are often caught in large numbers, are usually chilled with refrigerated or chilled seawater. Once chilled or frozen, the fish need further cooling to maintain the low temperature. There are key issues with fish cold store design and management, such as how large and energy efficient they are, and the way they are insulated and palletized.

An effective method of preserving the freshness of fish is to chill with ice by distributing ice uniformly around the fish. It is a safe cooling method that keeps the fish moist and in an easily stored form suitable for transport. It has become widely used since the development of mechanical refrigeration, which makes ice easy and cheap to produce. Ice is produced in various shapes; crushed ice and ice flakes, plates, tubes and blocks are commonly used to cool fish. Particularly effective is slurry ice, made from micro crystals of ice formed and suspended within a solution of water and a freezing point depressant, such as common salt.

A more recent development is pumpable ice technology. Pumpable ice flows like water, and because it is homogeneous, it cools fish faster than fresh water solid ice methods and eliminates freeze burns. It complies with HACCP and ISO food safety and public health standards, and uses less energy than conventional fresh water solid ice technologies.

### Control of Water Activity

The water activity, $a_w$, in a fish is defined as the ratio of the water vapour pressure in the flesh of the fish to the vapour pressure of pure water at the same temperature and pressure. It ranges between 0 and 1, and is a parameter that measures how available the water is in the flesh of the fish. Available water is necessary for the microbial and enzymatic reactions involved in spoilage. There are a number of techniques that have been or are used to tie up the available water or remove it by reducing the $a_w$. Traditionally, techniques such as drying, salting and smoking have been used, and have been used for thousands of years. These techniques can be very simple, for example, by using solar drying. In more recent times, freeze-drying, water binding humectants, and fully automated equipment with temperature and humidity control have been added. Often a combination of these techniques is used.

# 3

# Dried Fish

Fresh fish rapidly deteriorates unless some way can be found to preserve it. Drying is a method of food preservation that works by removing water from the food, which inhibits the growth of microorganisms.

Open air drying using sun and wind has been practiced since ancient times to preserve food. Water is usually removed by evaporation (air drying, sun drying, smoking or wind drying) but, in the case of freeze-drying, food is first frozen and then the water is removed by sublimation. Bacteria, yeasts and molds need the water in the food to grow, and drying effectively prevents them from surviving in the food.

Fish are preserved through such traditional methods as drying, smoking and salting. The oldest traditional way of preserving fish was to let the wind and sun dry it. Drying food is the world's oldest known preservation method, and dried fish has a storage life of several years. The method is cheap and effective in suitable climates; the work can be done by the fisherman and family, and the resulting product is easily transported to market.

## Types

### *Stockfish*

Stockfish is unsalted fish, especially cod, dried by cold air and wind on wooden racks on the foreshore. The drying racks are known as fish flakes. Cod is the most common fish used in stockfish production, though other whitefish, such as pollock, haddock, ling and tusk, are also used.

### Clipfish

Over the centuries, several variants of dried fish have evolved. Stockfish, dried as fresh fish and not salted, is often confused with clipfish, where the fish is salted before drying. After 2–3 weeks in salt the fish has saltmatured, and is transformed from wet salted fish to Clipfish through a drying process. The salted fish was earlier dried on rocks (clips) on the foreshore. The production method of Clipfish (or Bacalhau in Portuguese) was developed by the Portuguese who first mined salt near the brackish water of Aveiro, and brought it to Newfoundland where cod was available in massive quantities. (*q.v.*). Salting was not economically feasible until the 17th century, when cheap salt from southern Europe became available to the maritime nations of northern Europe.

Stockfish is cured in a process called fermentation where cold adapted bacteria matures the fish, similar to the maturing process of cheese. Clipfish is processed in a chemically curing process called saltmaturing, similar to the maturing processes of other saltmatured products like the Parma ham.

### Other

- Bacalhau is the Portuguese word for codfish and in a culinary context refers to dried and salted codfish. Fresh (unsalted) cod is referred to as *bacalhau fresco* (fresh cod). Bacalhau dishes are common in Portugal and Galicia, in the northwest of Spain, and to a lesser extent in former Portuguese colonies like Angola, Macau, and Brazil. There are said to be over 1000 recipes in Portugal alone and it can be considered the iconic ingredient of Portuguese cuisine (but curiously the only fish that is not consumed fresh in this fish-loving nation). It is often cooked on social occasions and is the Portuguese traditional Christmas dinner in some parts of Portugal.
- Baccalà is the name in the Venetian Language for salt cod. Most baccalà dishes require that the fish be soaked numerous times to remove excess saltiness.
- Balyk is the Russian term for the salted and dried soft parts of fish of large valuable species, such as sturgeon or salmon. Over time, the term has come to apply also to smoked fish of these species.
- Boknafisk is a variant of stockfish and is unsalted fish partially dried by sun and wind on drying flakes or on a wall. The most

common fish used for boknafisk is cod, but other types of fish can also be used. If herring is used, the dish is called *boknasild.*

- Dried squid
- Fesikh is a traditional Egyptian fish dish consisting of fermented salted and dried gray mullet, of the mugil family, a saltwater fish that lives in both the Mediterranean and the Red Seas. The traditional process of preparing it is to dry the fish in the sun before preserving it in salt.
- Gwamegi is a Korean half-dried Pacific herring or Pacific saury made during winter. It is mostly eaten in the region of North Gyeongsang Province such as Pohang, Uljin, and Yeongdeok where a large amount of the fish are harvested. Guryongpo Harbor in Pohang is the most famous. Fresh herring or saury is frozen at -10 degrees celsius and is placed outdoors in December to repeat freezing at night and defreezing in the day.

  The process continues until the water content of the fish drops to approximately 40%.
- Harðfiskur is the Icelandic term for wind-dried fish.
- Katsuobushi is the Japanese name for dried, fermented, and smoked skipjack tuna, sometimes referred to as bonito).
- A kipper is a whole herring, a small, oily fish, that has been split from tail to head, gutted, salted or pickled, and cold smoked.
- Kusaya is a Japanese style salted, dried and fermented fish. It has a pungent smell, similar to the fermented Swedish herring called surströmming.
- Maldive fish is cured tuna traditionally produced in the Maldives. It is a staple of the Maldivian cuisine, as well as Sri Lankan cuisine.
- Mojama (Spain) consists of filleted salt-cured tuna. The word *mojama* comes from the Arabic *musama* (dry), but its origins are Phoenician, specifically from *Gdr* (Gadir, Cádiz today), the first Phoenician settlement in the Western Mediterranean Sea. The Phoenicians had learned to dry tuna in sea salt so they could trade it. Mojama is made by curing tuna in salt for two days. The salt is then removed, the tuna is washed and then laid out to dry in the sun and the breeze (according to the traditional method) for fifteen to twenty days.

- Obambo is dried tilapia, prepared by cutting the fish open and drying it in the sun for several days. It is popular among the Luo and Luhya tribes, who live along the shores of Lake Victoria in Kenya. Traditionally, fishing was strictly forbidden during the rainy seasons, and people relied on obambo caught earlier and preserved.
- Tatami Iwashi is a Japanese processed food product made from baby sardines laid out and dried while entwined in a single layer to form a large mat-like sheet. Typically, this is done by drying them in the sun on a bamboo frame, a process that is evocative of the manufacture of traditional Japanese paper.

### Water Activity

The water activity, $a_w$, in a fish is defined as the ratio of the water vapour pressure in the flesh of the fish to the vapour pressure of pure water at the same temperature and pressure. It ranges between 0 and 1, and is a parameter that measures how available the water is in the flesh of the fish. Available water is necessary for the microbial and enzymatic reactions involved in spoilage. There are a number of techniques that have been or are used to tie up the available water or remove it by reducing the $a_w$. Traditionally, techniques such as drying, salting and smoking have been used, and have been used for thousands of years. These techniques can be very simple, for example, by using solar drying. In more recent times, freeze-drying, water binding humectants, and fully automated equipment with temperature and humidity control have been added. Often a combination of these techniques is used.

Salt cod has been produced for at least 500 years, since the time of the European discoveries of the New World. Before refrigeration, there was a need to preserve the codfish; drying and salting are ancient techniques to preserve nutrients and the process makes the codfish tastier.

The Portuguese tried to use this method of drying and salting on several varieties of fish from their waters, but the ideal fish came from much further north. With the "discovery" of Newfoundland in 1497, long after the Basque whalers arrived in Channel-Port aux Basques, they started fishing its cod-rich Grand Banks. Thus, *bacalhau* became a staple of the Portuguese cuisine, nicknamed *Fiel amigo* (faithful friend). From the 18th century, the town of Kristiansund in Norway became an important place of purchasing *bacalhau* or *klippfisk* (literally "cliff fish", since the fish was dried on stone cliffs by the sea to begin with.) Since the method was introduced by the Dutchman Jappe Ippes in abt 1690, the town had produced klippfisk and when the Spanish merchants arrived, it became a big industry. The bacalhau or bacalao dish is sometimes said to originate from Kristiansund, where it was introduced by the Spanish and Portuguese fish buyers and became very popular. Bacalao was common everyday food in north west Norway to this day, as it was cheap to make. In later years it is more eaten at special occasions. This dish was also popular in Portugal and other Roman Catholic countries, because of the many days (Fridays, Lent, and other festivals) on which the Church forbade the eating of meat. *Bacalhau* dishes were eaten instead.

## Salted Fish

Salted fish, such as kippered herring, is fish preserved or cured with salt. Drying and salting, either with dry salt or brine, was the only widely available method of preserving fish until the 19th century.

Salting is the preservation of food with dry edible salt. It is related to pickling (preparing food with brine, i.e. salty water), and is one of the oldest methods of preserving food. Salt inhibits the growth of microorganisms by drawing water out of microbial cells through osmosis. Concentrations of salt up to 20% are required to kill most species of unwanted bacteria. Smoking, often used in the process of curing meat, adds chemicals to the surface of meat that reduce the concentration of salt required. Salting is used because most bacteria, fungi and other potentially pathogenic organisms cannot survive in a highly salty environment, due to the hypertonic nature of salt. Any living cell in such an environment will become dehydrated through osmosis and die or become temporarily inactivated.

The water activity, $a_w$, in a fish is defined as the ratio of the water vapour pressure in the flesh of the fish to the vapour pressure of pure water at the same temperature and pressure. It ranges between 0 and 1, and is a parameter that measures how available the water is in the flesh of the fish. Available water is necessary for the microbial and enzymatic reactions involved in spoilage. There are a number of techniques that have been or are used to tie up the available water or remove it by reducing the $a_w$. Traditionally, techniques such as drying, salting and smoking have been used, and have been used for thousands of years. In more recent times, freeze-drying, water binding humectants, and fully automated equipment with temperature and humidity control have been added. Often a combination of these techniques is used.

## Smoked Fish

Smoked fish are fish that have been cured by smoking. Foods have been smoked by humans throughout history. Originally this was done as a preservative. In more recent times fish is readily preserved by refrigeration and freezing and the smoking of fish is generally done for the unique taste and flavour imparted by the smoking process.

### *Smoking Process*

According to Jeffrey J. Rozum "The process of smoking fish occurs through the use of fire. Wood contains three major components that are broken down in the burning process to form smoke. The burning process is called pyrolysis, which is simply defined as the chemical decomposition by heat. The major wood components are cellulose, hemicellulose and lignin."

"The major steps in the preparation of smoked fish are salting (bath or injection of liquid brine or dry salt mixture), cold smoking, cooling, packaging (air/vacuum or modified), and storage. Smoking, one of the oldest preservation methods, combines the effects of salting, drying, heating and smoking. Typical smoking of fish is either cold (28–32°C) or hot (70–80°C). Cold smoking does not cook the flesh, coagulate the proteins, inactivate food spoilage enzymes, or eliminate the food pathogens, and hence refrigerated storage is necessary until consumption."

### *Smokehouses*

A smokehouse is a building where fish or meat is cured with smoke. In a traditional fishing village, a smokehouse was often attached to a fisherman's cottage. The smoked products might be stored in the building, sometimes for a year or more. Traditional smokehouses served both as smokers and to store the smoked fish. Fish could be preserved if it was cured with salt and cold smoked for two weeks or longer. Smokehouses were often secured to prevent animals and thieves from accessing the food.

### *Traditional versus Mechanical*

Today there are two main methods of smoking fish: The traditional method and the mechanical method. The traditional method involves the fish being suspended in smokehouses over slowly smouldering wood shavings. The fish are left overnight to be naturally infused with smoke.

In the mechanical method smoke is generated through the use of smoke condensates, which are created by the industrial process of turning smoke into a solid or liquid form. The flow of smoke in the mechanical kiln is computer controlled and the fish generally spend less time being smoked than in a traditional kiln.

Laminar air-flow technology allows mechanical kilns to achieve a higher production rate, while the use of micro-processors has allowed mechanical kiln smokers increased sensor coverage within the kiln. However, traditional smokers argue that this removes the human element from production. They feel that a computer is no substitute for many years of hands–on experience. Most traditional smokehouses have developed their method across generations. Mechanical kiln smoked fish represents a quantitative approach where supermarkets represent the main market. Traditional fish smoking is a qualitative process. Traditional smoked fish is a high–end product sought after by restaurants.

The most common types of smoked fish in the US are salmon, mackerel, whitefish and trout, although other smoked fish is also available regionally or from many ethnic stores. Salmon, mackerel and herring are universally available both hot-smoked and cold-smoked, while most other fish is traditionally preserved by only one of the smoking methods. A common name for cold-smoked salmon is *lox,* of which many different types are available, usually identified by point of origin (e.g., from Scotland, Norway, Holland, the Pacific, and Nova Scotia, Canada—the latter usually identified as *Nova lox* or just *Nova*). Traditionally, *lox* designates brined rather than smoked salmon, but the linguistic boundary between the two types of products has long been erased. *Gravad lax* or gravlax remains the only type that is unmistakably *not* smoked fish. However, commercial labels still identify most smoked products as “smoked salmon” rather than “lox”.

Most other smoked fish in the US is hot-smoked, although cold-smoked mackerel is always available in East-European delis, along with cold-smoked sturgeon, sea bass, halibut or turbot and many other varieties. Jewish delis often sell, in addition to lox, hot-smoked whitefish, mackerel, trout, and sablefish (also sometimes referred to as black cod in its fresh state). Along the Mississippi River, hot-smoked locally caught sturgeon is also available. Traditionally, in the US, cold-smoked fish, other than salmon, is considered “raw” and thus unsafe to consume without cooking. For this reason, in the US, cold-smoked fish is largely confined to speciality and ethnic shops.

In the Netherlands, commonly available varieties include both hot- and cold-smoked mackerel, herring and Baltic sprats. Hot-smoked eel is a speciality in the Northern provinces, but is a popular deli item throughout the country. Smoked fish is a prominent item in Russian cuisine, Ashkenazi Jewish Cuisine, and Scandinavian cuisine, as well as several Eastern and Central European cuisines and the Pacific Northwest cuisine.

English, Scottish and Canadian cuisine incorporate a variety of strongly brined, smoked herring that used to be known as "red herring". With the increased use of the idiomatic expression "red herring", references to the smoked fish product in this manner declined. A more common contemporary name for it is kippers, or kippered herring. Kippered herring traditionally undergoes further processing (soaking and cooking) before consumption. Arbroath Smokies (haddock) and Traditional Grimsby smoked fish (haddock and cod) have both received Protected Geographical Indication status from the European Commission, which restricts use of the name to fish that is processed using specific methods within a defined geographical area. Other smoked fish products from the UK include Finnan Haddie and Bloater.

### Physical Control of Microbial Loads

Heat or ionizing irradiation can be used to kill the bacteria that cause decomposition. Heat is applied by cooking, blanching or microwave heating in a manner that pasteurizes or sterilizes fish products. Cooking or pasteurizing does not completely inactivate microorganisms and may need to be followed with refrigeration to preserve fish products and increase their shelf life. Sterilised products are stable at ambient temperatures up to 40°C, but to ensure they remain sterilized they need packaging in metal cans or retortable pouches before the heat treatment.

### Chemical Control of Microbial Loads

Microbial growth and proliferation can be inhibited by a technique called biopreservation. Biopreservation is achieved by adding antimicrobials or by increasing the acidity of the fish muscle. Most bacteria stop multiplying when the pH is less than 4.5. Acidity is increased by fermentation, marination or by directly adding acids (acetic, citric, lactic) to fish products. Lactic acid bacteria produce the antimicrobial nisin which further enhances preservation. Other preservatives include nitrites, sulphites, sorbates, benzoates and essential oils.

### Control of the Oxygen Reduction Potential

Spoilage bacteria and lipid oxidation usually need oxygen, so reducing the oxygen around fish can increase shelf life. This is done by controlling or modifying the atmosphere around the fish, or by vacuum packaging. Controlled or modified atmospheres have specific combinations of oxygen, carbon dioxide and nitrogen, and the method is often combined with refrigeration for more effective fish preservation.

### Combined Techniques

Two or more of these techniques are often combined. This can improve preservation and reduce unwanted side effects such as the denaturation of nutrients by severe heat treatments. Common combinations are salting/drying, salting/marinating, salting/smoking, drying/smoking, pasteurization/refrigeration and controlled atmosphere/refrigeration. Other process combinations are currently being developed along the multiple hurdle theory.

### Hurdle Technology

Hurdle technology is a method of ensuring that pathogens in food products can be eliminated or controlled. This means the food products will be safe for consumption, and their shelf life will be extended. Hurdle technology usually works by combining more than one approach. These approaches can be thought of as "hurdles" the pathogen has to overcome if it is to remain active in the food. The right combination of hurdles can ensure all pathogens are eliminated or rendered harmless in the final product.

Hurdle technology has been defined by Leistner (2000) as an intelligent combination of hurdles which secures the microbial safety and stability as well as the organoleptic and nutritional quality and the economic viability of food products. The organoleptic quality of the food refers to its sensory properties, that is its look, taste, smell and texture.

Examples of hurdles in a food system are high temperature during processing, low temperature during storage, increasing the acidity, lowering the water activity or redox potential, or the presence of preservatives. According to the type of pathogens and how risky they are, the intensity of the hurdles can be adjusted individually to meet consumer preferences in an economical way, without compromising the safety of the product.

## Automated Processes

"The search for higher productivity and the increase of labour cost has driven the development of computer vision technology, electronic scales and automatic skinning and filleting machines."

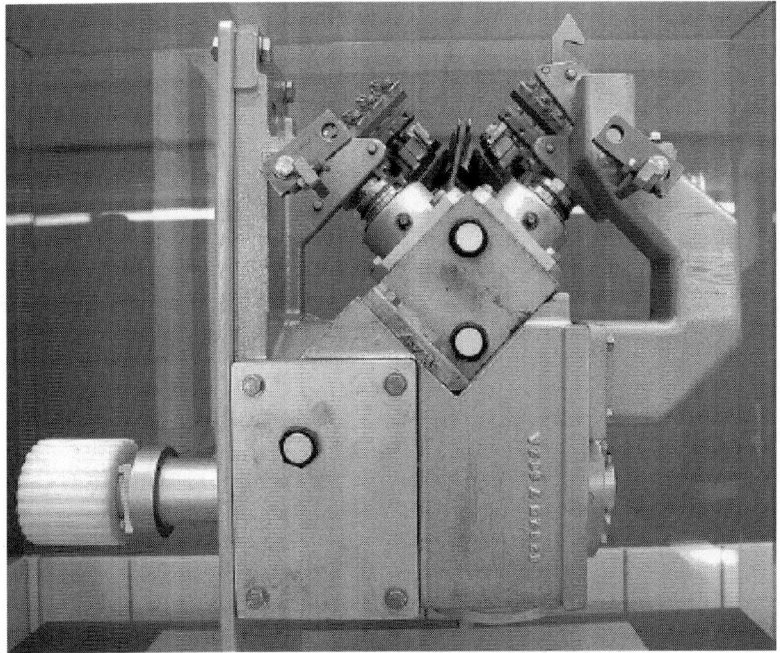

***Figure:*** *Automatic knives for filleting fish*

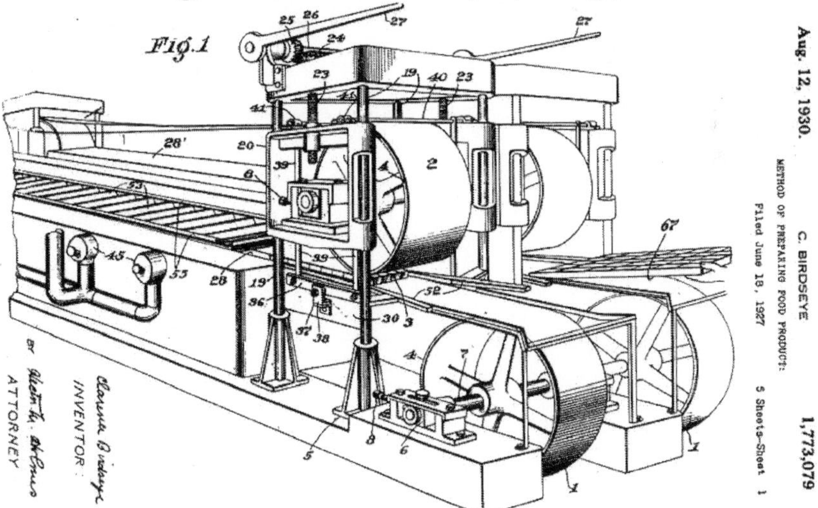


***Figure:*** *Patent issued to Clarence Birdseye for the production of quick-frozen fish, 1930*

## Waste Management

Waste produced during fish processing operations can be solid or liquid.

- Solid wastes: include skin, viscera, fish heads and carcasses (fish bones). Solid waste can be recycled in fish meal plants or it can be treated as municipal waste.
- Liquid wastes: include bloodwater and brine from drained storage tanks, and water discharges from washing and cleaning. This waste may need holding temporarily, and should be disposed of without damage to the environment. How liquid waste should be disposed from fish processing operations depends on the content levels in the waste of solid and organic matter, as well as nitrogen and phosphorus content, and oil and grease content. It also depends on an assessment of parameters such acidity levels, temperature, odour, and biochemical oxygen demand and chemical oxygen demand. The magnitude of waste management issues depends on how much waste volume there is, the nature of the pollutants it carries, the rate at which it is discharged and the capacity of the receiving environment to assimilate the pollutants. Many countries dispose of such liquid wastes through their municipal sewage systems or directly into a waterway. The receiving waterbody should be able to degrade the organic and inorganic waste components in a way that does not damage the aquatic ecosystem.

Treatments can be primary and secondary.

- Primary treatments: use physical methods such as flotation, screening, and sedimentation to remove oil and grease and other suspended solids.
- Secondary treatments: use biological and physicochemical means. Biological treatments use microorganisms to metabolise the organic polluting matter into energy and biomass. "These microorganisms can be aerobic or anaerobic. The most used aerobic processes are activated sludge system, aerated lagoons, trickling filters or bacterial beds and the rotating biological contractors. In anaerobic processes, the anaerobic microorganisms digest the organic matter in tanks to produce gases (mainly methane and $CO_2$) and biomass. Anaerobic digesters are sometimes heated, using part of the methane produced, to maintain a temperature of 30 to 35°C. In the physicochemical treatments, also called coagulation-flocculation, a chemical substance is added to the effluent to reduce the surface charges responsible for particle repulsions in a colloidal suspension, thus reducing the forces that keep its particles apart. This reduction in charge causes flocculation (agglomeration) and

particles of larger sizes are settled and clarified effluent is obtained. The sludge produced by primary and secondary treatments is further processed in digesting tanks through anaerobic processes or sprayed over land as a fertilizer. In the latter case, care must be exercised to ensure that the sludge is freed of its pathogens."

### Transport

Fish is transported widely in ships, and by land and air, and much fish is traded internationally. It is traded live, fresh, frozen, cured and canned. Live, fresh and frozen fish need special care.

- *Live fish:* When live fish are transported they need oxygen, and the carbon dioxide and ammonia that result from respiration must not be allowed to build up. Most fish transported live are placed in water supersaturated with oxygen (though catfish can breathe air directly through their gills and body skin, and the climbing perch has special air-breathing organs). The fish are often "conditioned" (starved) before they are transported to reduce their metabolism and increase packing density, and the water can be cooled to further reduce metabolism. Live crustaceans can be packed in wet sawdust to keep the air humid.
- *By air:* Over five percent of the global fish production is transported by air. Air transport needs special care in preparation and handling and careful scheduling. Airline transport hubs often require cargo transfers under their own tight schedules. This can influence when the product is delivered, and consequently the condition it is in when it is delivered. The air shipment of leaking seafood packages causes corrosion damage to aircraft, and each year, in the US, requires millions of dollars to repair the damage. Most airlines prefer fish that is packed in dry ice or gel, and not packed in ice.
- *By land or sea:* "The most challenging aspect of fish transportation by sea or by road is the maintenance of the cold chain, for fresh, chilled and frozen products and the optimisation of the packing and stowage density. Maintaining the cold chain requires the use of insulated containers or transport vehicles and adequate quantities of coolants or mechanical refrigeration. Continuous temperature monitors are used to provide evidence that the cold chain has not been broken during transportation. Excellent development in food packaging and handling allow

rapid and efficient loading, transport and unloading of fish and fishery products by road or by sea. Also, transport of fish by sea allows for the use of special containers that carry fish under vacuum, modified or controlled atmosphere, combined with refrigeration."

### Quality and Safety

The International Organisation for Standardisation, ISO, is the worldwide federation of national standards bodies. ISO defines *quality* as "the totality of features and characteristics of a product or service that bear on its ability to satisfy stated or implied needs."(ISO 8402). The quality of fish and fish products depends on safe and hygienic practices. Outbreaks of fish-borne illnesses are reduced if appropriate practices are followed when handling, manufacturing, refrigerating and transporting fish and fish products. Ensuring standards of quality and safety are high also minimizes the post-harvest losses."

"The fishing industry must ensure that their fish handling, processing and transportation facilities meet requisite standards. Adequate training of both industry and control authority staff must be provided by support institutions, and channels for feedback from consumers established. Ensuring high standards for quality and safety is good economics, minimizing losses that result from spoilage, damage to trade and from illness among consumers."

Fish processing highly involves very strict controls and measurements in order to ensure that all processing stages have been carried out hygienically. Thus, all fish processing companies are highly recommended to join a certain type of food safety system. One of the certifications that are commonly known is the Hazard Analysis Critical Control Points (HACCP).

### Hazard Analysis and Critical Control Points

HACCP is a system which identifies hazards and implements measures for their control. It was first developed in 1960 by NASA to ensure food safety for the manned space program. The main objectives of NASA were to prevent food safety problems and control food borne diseases. HACCP has been widely used by food industry since the late 1970 and now it is internationally recognised as the best system for ensuring food safety.

"The Hazard Analysis and Critical Control Points (HACCP) system of assuring food safety and quality has now gained worldwide recognition as the most cost-effective and reliable system available.

It is based on the identification of risks, minimizing those risks through the design and layout of the physical environment in which high standards of hygiene can be assured, sets measurable standards and establishes monitoring systems. HACCP also establishes procedures for verifying that the system is working effectively. HACCP is a sufficiently flexible system to be successfully applied at all critical stages — from harvesting of fish to reaching the consumer. For such a system to work successfully, all stakeholders must cooperate which entails increasing the national capacity for introducing and maintaining HACCP measures. The system's control authority needs to design and implement the system, ensuring that monitoring and corrective measures are put in place."

HACCP is endorsed by the:

- FAO (Food and Agriculture Organisation)
- Codex Alimentarius (a commission of the United Nations)
- FDA (US Food and Drug Administration)
- European Union
- WHO (World Health Organisation)

There are seven basic principles:

- Principle 1: Conduct a hazard analysis.
- Principle 2: After assessing all the processing steps, the Critical control point (CCP) is controlled. CCP are points which determine and control significant hazards in a food manufacturing process.
- Principle 3: Set up critical limits in order to ensure that the hazard identified is being controlled effectively.
- Principle 4: Establish a system so as to monitor the CCP.
- Principle 5: Establish corrective actions where the critical limit has not been met. Appropriate actions need to be taken which can be on a short or long-term basis. All records must be sustained accurately.
- Principle 6: Establish authentication procedures so as to confirm if the principles imposed by HACCP documents are being respected effectively and all records are being taken.
- Principle 7: Analyze if the HACCP plan are working effectively.

## Final Products

Finfish, or parts of finfish, are typically presented physically for marketing in one of the following forms

- whole fish: the fish as it originally came from the water, with no physical processing
- drawn fish: a whole fish which has been eviscerated, that is, had its internal organs removed
- dressed fish: fish that has been scaled and eviscerated, and is ready to cook.
- pan dressed fish: a dressed fish which has had its head, tail, and fins removed, so it will fit in a pan.
- filleted fish: the "fleshy sides of the fish, cut lengthwise from the fish along the backbone. They are usually boneless, although in some fish small bones called "pins" may be present; skin may be present on one side, too. Butterfly fillets may be available. This refers to two fillets held together by the uncut flesh and skin of the belly"
- fish steaks: large dressed fish can be cut into cross section slices, usually half to one inch thick, and usually with a cross section of the backbone
- fish sticks: "are pieces of fish cut from blocks of frozen fillets into portions at least 3/8-inch thick. Sticks are available in fried form ready to heat or frozen raw, coated with batter and breaded, ready to be cooked"
- fish cakes: are "prepared from flaked fish, potatoes, and seasonings, and shaped into cakes, coated with batter, breaded, and then packaged and frozen, ready-to-be-cooked"
- fish fingers
- fish roe

## *Value Addition*

In general value addition means "any additional activity that in one way or the other change the nature of a product thus adding to its value at the time of sale." Value addition is an expanding sector in the food processing industry, especially in export markets. Value is added to fish and fishery products depending on the requirement of different markets. Globally a transition period is taking place where cooked products are replacing traditional raw products in consumer preference.

"In addition to preservation, fish can be industrially processed into a wide array of products to increase their economic value and allow the fishing industry and exporting countries to reap the full benefits of their aquatic resources. In addition, value processes generate further employment and hard currency earnings. This is more important nowadays because of societal changes that have led to the development of outdoor catering, convenience products and food services requiring fish products ready to eat or requiring little preparation before serving."

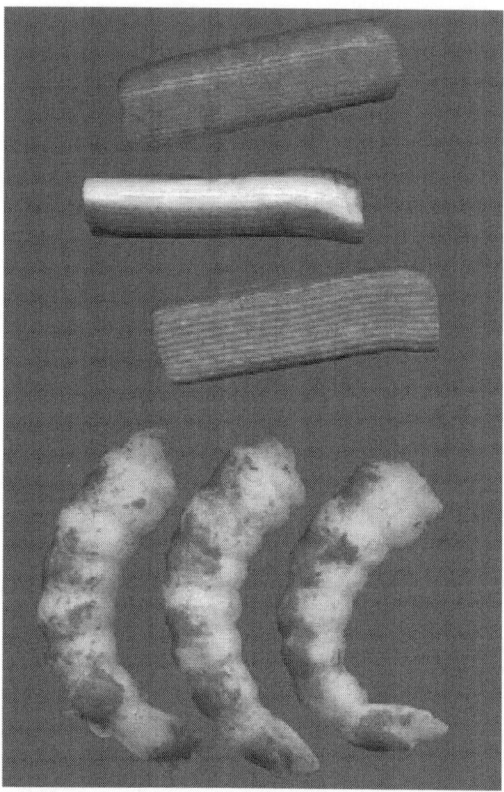

***Figure:*** *Imitation crab and imitation shrimp made from surimi*

"However, despite the availability of technology, careful consideration should be given to the economic feasibility aspects, including distribution, marketing, quality assurance and trade barriers, before embarking on a value addition fish process."

- Surimi: Surimi and surimi-based products are an example of value added products. Surimi is prepared from the mechanically deboned, washed (bleached) and stabilised flesh of fish. "It is an intermediate product used in the preparation of a variety of ready to eat seafood such as kamaboko, fish sausage, crab legs and imitation shrimp products. Surimi-based products are gaining more prominence worldwide, because of the emergence of Japanese restaurants and culinary traditions in North America, Europe and elsewhere. Ideally, surimi should be made from low-value, white fish with excellent gelling ability and which are abundant and available year-round. At present, Alaskan pollack accounts for a large proportion of the surimi supply. Other species, such as sardine, mackerel, barracuda, striped mullet have been successfully used for surimi production."

- Fishmeal and fish oil: "A significant proportion of the world catch (20 percent) is processed into fishmeal and fish oil. Fishmeal is a ground solid product that is obtained by removing most of the water and some or all of the oil from fish or fish waste. This industry was launched in the 19th century, based mainly on surplus catches of herring from seasonal coastal fisheries to produce oil for industrial uses in leather tanning and in the production of soap, glycerol and other non-food products.

  Presently, it uses small oily fish to produce fishmeal and oil. It is worthy to mention that, only where it is uneconomic or impracticable for human consumption, should the catch be reduced to fishmeal and oil. Indeed, cycling fish through poultry or pigs is a loss because there is a need for 3 kg of edible fish to produce approximately 1 kg of edible chicken or pork."

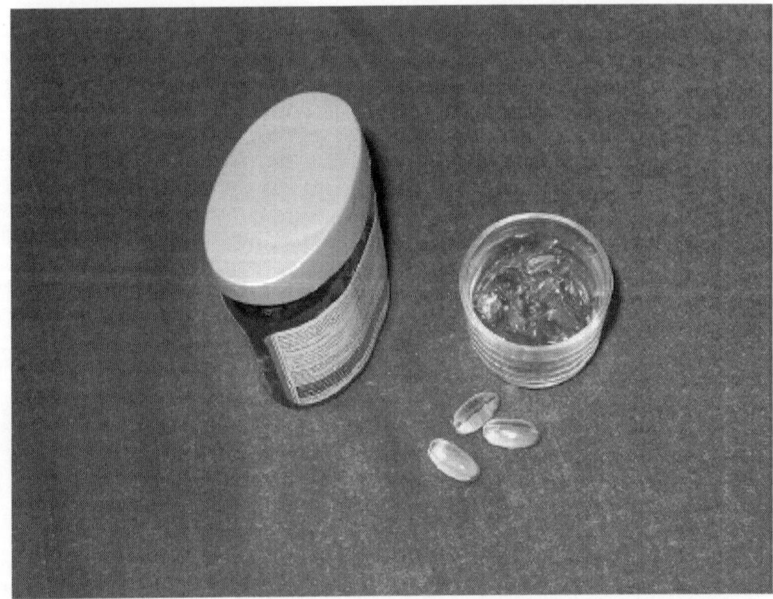

***Figure:*** *Fish oil capsules*

## History

There is evidence humans have been processing fish since the early Holocene. For example, fishbones (c. 8140–7550 BP, uncalibrated) at Atlit-Yam, a submerged Neolithic site off Israel, have been analysed. What emerged was a picture of "a pile of fish gutted and processed in a size-dependent manner, and then stored for future consumption or trade.

This scenario suggests that technology for fish storage was already available, and that the Atlit-Yam inhabitants could enjoy the economic stability resulting from food storage and trade with mainland sites."

## Fish Products

Fisheries are estimated to currently provide 16% of the world population's protein. The flesh of many fish are primarily valued as a source of food; there are many edible species of fish. Other marine life taken as food includes shellfish, crustaceans, sea cucumber, jellyfish and roe.

Fish and other marine life are also be used for many other uses: pearls and mother-of-pearl, sharkskin and rayskin. Sea horses, star fish, sea urchins and sea cucumber are used in traditional Chinese medicine.

Tyrian purple is a pigment made from marine snails, sepia is a pigment made from the inky secretions of cuttlefish. Fish glue has long been valued for its use in all manner of products. Isinglass is used for the clarification of wine and beer. Fish emulsion is a fertilizer emulsion that is produced from the fluid remains of fish processed for fish oil and fish meal.

In the industry the term *seafood products* is often used instead of *fish products.*

## Fish Marketing

Fish marketing, is the marketing and sale of fish products. Fish markets are marketplace used for the trade in and sale of fish and other seafood. They can be dedicated to wholesale trade between fishermen and fish merchants, or to the sale of seafood to individual consumers, or to both. Retail fish markets, a type of wet market, often sell street food as well.

Most shrimps are sold frozen and are marketed in different categories. The live food fish trade is a global system that links fishing communities with markets.

### Live Fish Trade

The live fish trade can refer to the live food fish trade (for dinner) or to the ornamental fish trade (for aquariums). The fish can come from many places, but most comes from Southeast Asia. The live food fish trade is a global system that links fishing communities with markets, primarily in Hong Kong and mainland China. Many of the fish are captured on coral reefs in Southeast Asia or the Pacific Island nations.

### Consumer Demand

Within the live food trade there are certain types of fish demanded more often by consumers, particularly smaller and medium-sized fish. According to the book While Stocks Last: The Live Reef Food Fish Trade consumer demand has caused the fish captured on coral reefs to be the most valued fish in the trade. Consumers are important because they are directly purchasing these fish species at restaurants and stores. In addition to these types of fishes, many juvenile fish are used for the live food trade. There are also cultural and regional preferences among consumers, for example, Chinese consumers often prefer their fish to be reddish in colour believing the colour to be auspicious (Sadovy, Y.J, 393). These preferences inevitably affect the biodiversity of marine life making certain fish species rarer to find.

The life fish food trade is a lucrative business. According to University of Washington Professor Patrick Christie, live fish caught for food export earns approximately $6000 a ton. To help support themselves and their families, fishermen in Oceania and Southeast Asia sometimes use illegal fishing methods. Although many feel the fish are worth the cost, a typical dinner can cost up to one hundred dollars per kilogram. The wholesale value on these fish is anywhere from eleven US dollars to sixty-three US dollars per kilogram, meaning there's a large markup and resale value. (Hong Kong alone is estimated to be about four hundred million US dollars a year.) Because this trade frequently uses illegal methods of collecting (using cyanide), there is no way to know for sure how much money is being made each year on live fish trade, although estimates conclude probably over one billion US dollars each year.

As is often the case, consumers are willing to pay large amounts of money on rare and fresh fish. One 500-pound, polka-dot grouper, estimated to be more than a century old, was hacked into fillets by seven kitchen workers in about half an hour, the Economist reports. It was expected to bring about $15,000. (Moll 1996)

## Market and Trade Routes

The centre for the Live Food Fish Trade is located in Hong Kong — the markets consumers contribute $400 million to the estimated $1 billion of the trades global value (Seaweb). Total imports flowing into Hong Kong included 10153 metric tons, of which 30 percent was re-exported to mainland China (Montaldi). Other major markets include Singapore, mainland China, and Taiwan (Graham). The primary suppliers of wild caught fish are Indonesia (accounting for nearly 50 percent of Hong Kong's imports), Thailand, Malaysia, Australia, and Vietnam (Graham). However, Taiwan and Malaysia are leading the charge towards farmed live fish specialising in an industry that "harvested annually has probably been in the billions [Metric Tons](Graham)." Farming live fish is gaining popularity as tastes for live fish are burgeoning across South Asia and countries look to become more and more self sustainable, this is appareant in nations with sizable Chinese populations such as Indonesia and Malaysia.

Hong Kong and China are the dominant markets for the live fish, in addition to other cities in the region that have large Chinese populations, including Singapore and Kuala Lumpur (BusinessWorld Nov. 1999). In Southeast Asia, Singapore alone consumes 500 tons of live coral fish a year(ibid). Exports from Southeast Asia rose to over 5,000 tons in 1995 from 400 tons in 1989(ibid). However, in 1996, exports declined by 22%(ibid). Indonesia, which accounts for over 60% of the harvest, saw exports falling by over 450 tons(ibid). That same year, other Southeast Asian countries have experienced similar drops in stocks of live coral fish for food(ibid). In 1996, the Philippines' exports were halved, while Malaysian exports declined by over 30%(ibid). These decreases in catch have been due to the excessive amount of fish caught for exports and the degradation of the coral reefs from such procedures.

## Corruption in the Trade

The live fish trade is a complex issue that involves many different perspectives, all of which must be considered in trying to approach a solution. While, at first, one may point the finger at the fishermen themselves as the criminals, there are many other factors. One is the economic disparity of many of the communities that take part in cyanide, dynamite, or other illegal fishing practices. 40% of the Filipino population and 27% of the Indonesian population is considered to be in poverty. The World Factbook)] Many whose livelihood once depended on fishing or agriculture are realising it is more lucrative to participate in illegal fishing activities.

Community members who are not a part of the trade are affected by the activities of these illegal fishers. The cyanide fishers profit by taking away from everyone else's trade and food. (Lowe, 7) '*If people were using poison and my take dropped to only a little, I would accept it,*' Puah said. '*But I feel heartsick...I catch nothing at all. I have not caught a big fish in a month so there's no point in going fishing this afternoon.*'" (Lowe, 7) The local people are often helpless to protect themselves, as government and law enforcement officials have "open pockets" and are also involved in the trade by turning a blind eye to the illegal actions and receiving a take of the profits. "Culpability in cyanide use cannot be understood apart from the larger structures of corruption that permeates resource extraction throughout Indonesia. The Indonesian State bureaucracy extends from Jakarta down to the village level, and radiates out into villages through kinship connections. It is the factor most tightly correlated with illegal trade in natural resources throughout Indonesia."

### *Cause and Rffect*

Coral reefs found in the South Pacific are regarded as the "rainforest of the sea" harboring countless fish species large and small. However, recently the live fish trade has threatened the sanctity of these endangered areas. The Global Coral Reef Monitoring Network has issued a recent report that estimates that 25% of the world's reefs are severely damaged and another third are in grave danger. (Moore Online) The live fish trade is part of this alarming ecological trend caused by the popular use of cyanide which is injected into the coral reefs to stun inhabiting fish so they can be easily caught by nets. It is estimated that since the 1960s, more than one million kilograms of cyanide has been squirted into Philippine reefs alone, and since then the practice has spread throughout the South Pacific. (Moore Online) The live fish trade is only growing, in 1994 the Philippines exported 200,000 kg of live fish; by 2004 the Philippines were annually exporting 800,000 kg annually. (Aguiba Online) Although Asian markets are the primary buyers of live reef fish for food, the recently created U.S. Coral Reef Task Force has concluded that the U.S. is the primary purchaser of live reef fish for aquariums as well as eclectic jewellery. (Moore Online) Even though the use of cyanide in the live fish trade is severely detrimental, one must realise that this issue is multidimensional. Small-scale native fishermen of the small South Pacific coastal communities are the backbone of the live fish trade, and are forced to resort to the illegal use of sodium cyanide due to demand and high prices offered by the industry.

## Fishing Techniques

While live food fish trade can be very profitable for those involved, there are many dangerous aspects to it. Through the use of illegal practices such as cyanide fishing, coral reefs and fish communities are put in grave danger. The process of cyanide fishing involves injecting crushed cyanide tablets and squirting this solution from a bottle toward the targeted fish on top of coral heads. Specifically, the cyanide kills coral polyps, symbiotic algae, and other coral reefs organisms that are necessary for maintaining the health of the coral reef. These damages eventually deteriorate the coral reef and lead it into collapse of the entire coral reef ecosystem.

The effect on the targeted fish is disorientation and semi-paralysis. After being squirted with cyanide the fish is easy brought to the surface and kept alive in small, on board container. Fishermen often understand that this practice is harmful and will offer locals a portion of the fish in order to continue fishing. When ingested, small levels of cyanide accumulate in the system causing weakness of fingers and toes, failure of the thyroid gland and blurred vision. Divers without experience may come in direct contact with cyanide, causing death. Estimates show that since the 1960s, over a million kilograms of cyanide have been squirted into the coral reefs of the Philippines alone (Bryant et al.).

The harm upon the reefs is coming full circle and having a social impact through the limited fish stocks. As fish are depleted from these fishing techniques, the fishermen are having a more difficult time feeding themselves (Cyanide). Additionally, the use of explosives may be used as a fishing technique in the live food fish trade. While the majority of these fish do not survive the blast of such explosions, the remaining fish that are only stunned are collected for the live food fish trade. The use of cyanide makes a stronger argument in that the more coral reef fish captured alive, the more lucrative the catch is for the fisherman. Live fish, according to the [WWF], fetch five times more than a dead fish (Cyanide).

This is why banning the live-fish trade would similarly harm the reefs. Fisherman would resort to the "dead fish trade", be forced to deal in a larger quantity of fish, and the process of dangerous fishing techniques continues. To illustrate the effects of repeated explosions in the proximity of a coral reef system, roughly one half of the coral reefs in Komodo National Park in Indonesia have been destroyed (Moore). The use of cyanide and explosives in fishing proves to be an effective technique in catching fish, but its powers are indiscriminating and as a result the coral reefs are being held hostage to such practices. The future of the reefs are in question just as are the futures of those who subsist from them because "from a long term perspective, the question of ethics of using the ocean...contains a commitment for future generations"

### *Impacts on Humans*

In communities like those in the Philippines and Indonesia, people are participating in the live food fish trade because it is a source of income, or at least a source of temporary income. For some communities this is one of the few income-generating opportunities. Along with the environmental, ecological, and economic consequences of this industry, there are serious health risks as well. Because of inadequate training and lack of quality equipment, divers, especially young men are in large risk of paralysis.

### *Sustainable Practices*

Because of the great profitability of this industry, there is a great incentive to identify sustainable practices. The Marine Aquarium Council (MAC) works to offer the hobbyist with a product that is certified as environmentally sound and sustainable. Additionally, the International Marinelife Alliance (IMA), The Nature Conservancy (TNC), and MAC are working with the Hong Kong Chamber of Seafood Merchants to develop standards for the live fish trade. The Hong Kong Seafood Merchants represent ninety percent of the buyers of live reef food fish in Hong Kong and have an extensive impact on collection practices.

In an effort to address the damage inflicted on coral reef ecosystems and fish stocks, aquaculture is being utilised to reduce pressure on coral reefs. However, initial efforts to farm grouper have met with significant challenges. There are difficulties with fragile grouper seed that can make it more expensive than wild caught larvae, which can affect natural replenishment rates. Additionally, there are problems with finding suitable food, disease and cannibalism (Johannes and Ogburn). Efforts are also being made in regards to the aquarium fish trade. Juvenile fish are being captured and raised specifically for the industry. However, there are debates as to whether this practice will affect replenishment rates. "The age of the juveniles is pivotal to the debate, harvesting of postlarvae from the water column is considered to have a much lower (negligible) impact on rates of replenishment than the removal of the larger juveniles from benthic habitats because the postlarvae have yet to undergo severe mortality" (Bell, Doherty and Hair). If studies determine that the capture of juveniles is sustainable, it may help in mitigating the damage from cyanide fishing.

It should also be noted that aquaculture production, specifically grouper rearing is rapidly expanding in Asia. From 1998 to 2001 the Indo-Pacific countries involved in aquaculture; China, Indonesia, Republic of Korea, Kuwait, Malaysia, Philippines, Singapore, and Thailand witnessed a 119 percent increase in output (FAO fishery Information Data and Statistics Unit 2003). The explosion in this practice can most likely be attributed to the large profit margins that can be derived in very little time. It is estimated that the majority of farms after annual returns can be paid back in less than one year (Siar et al. 2002). In comparison to other species of fish such as the Milkfish, the Grouper, because of high demand is able to garner high rates of return, in order to earn 1,000 dollars a grouper farm would only have to raise 400 kilograms in contrast to 5,000 kilograms of Milkfish (Ibid).

## Shrimp Marketing

Shrimp are marketed and commercialised with several issues in mind. Most shrimp are sold frozen and marketed based on their categorisation of presentation, grading, colour, and uniformity. Most shrimps are sold frozen and are marketed in different categories; the main factors for categorization are presentation, grading, colour, and uniformity.

### *Presentation*

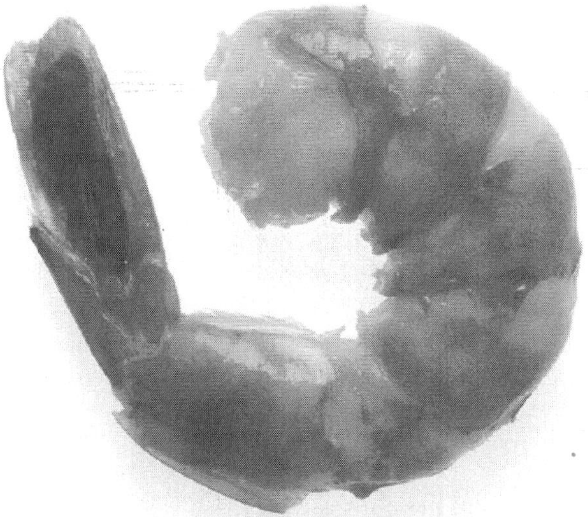

***Figure:*** *A steamed tail-on shrimp.*

The main forms of presentation are head on shell on (HOSO), shell on (SO or "green headless shrimp"), peeled tail on (PTO), peeled undeveined (PUD), peeled and deveined (P&D), and butterfly tail on (BTTY-TO). Sometimes a letter 'F' is placed in front of these abbreviation for the presentation in order to state that the shrimp comes from a farm (example: FSO – farmed, shell on). European and Asian markets prefer the HOSO presentation (which is a whole shrimp), while the American shrimp market prefers the remaining presentations.

### Grading

Shrimp are graded according to their count per weight. HOSO shrimps are graded in units per kilogram (30/40, 40/50, 50/60, etc. pcs/kg). The standard pack is in a 2 kg box, 10 boxes into a master carton. The remaining presentations are graded in units per pound (U15, 16/20, 21/25, 26/30, 31/35, 36/40, 41/50, etc. pcs/lb). The standard pack is in a 5 lb box, 10 boxes into a master carton.

The numbers in the grading code indicate maximum and minimum quantity of pieces per unit weight, with U standing for "under".

### Colour

HOSO shrimp are also graded according to their colour, with A1 being the lightest colour and A5 the darkest colour. Shrimp tend to take on the colour of their habitat, so sandy ponds tend to yield an A1 colour shrimp, and shrimp ponds with plastic black liners tend to yield A5 colour shrimp. A2–A3 colour shrimp are preferred for fresh commercialisation, and A4–A5 colour shrimp are preferred for cooked commercialisation. The other presentations are not usually graded by colour.

### Uniformity

Another concept to grade is the uniformity. This is measured by visually selecting the ten largest and the ten smallest pieces from 1 kg of product. The two groups are then weighed separately and the weight of the large pieces group is divided by the weight of the small pieces group. The result is the uniformity factor. Normally the maximum accepted value for the uniformity factor is 1.5.

### Fish Markets

A fish market is a marketplace used for marketing fish products. It can be dedicated to wholesale trade between fishermen and fish merchants, or to the sale of seafood to individual consumers, or to both. Retail fish markets, a type of wet market, often sell street food as well. Fish markets range in size from small fish stalls, such as the one in the photo at the right, to the great Tsukiji fish market in Tokyo, turning over about 660,000 tonnes a year.

The term *fish market* can refer to the process of fish marketing in general, but this article is concerned with physical marketplaces. There is a long history of fish markets from the time of ancient Greece. They served as a public space where large numbers of people could gather and discuss current events and local politics. Because seafood is quick to spoil, fish markets are historically most often found in seaside towns. Once ice or other simple cooling methods became available, some were also established in large inland cities that had good trade routes to the coast.

Since refrigeration and rapid transport became available in the 19th and 20th century, fish markets can technically be established at any place. However, because modern trade logistics in general has shifted away from marketplaces and towards retail outlets, such as supermarkets, most seafood worldwide is now sold to consumers through these venues, like most other foodstuffs.

Consequently, most major fish markets now mainly deal with wholesale trade, and the existing major fish retail markets continue to operate as much for traditional reasons as for commercial ones. Both types of fish markets are often tourist attractions as well.

### Traditional Sector

The traditional fishing industry, or artisan fishing, are terms used to describe small scale commercial or subsistence fishing practises, particularly using traditional techniques such as rod and tackle, arrows and harpoons, throw nets and drag nets, etc. It does not usually cover the concept of fishing for sport, and might be used when talking about the pressures between large scale modern commercial fishing practises and traditional methods, or when aid programs are targeted specifically at fishing at or near subsistence levels.

### Recreational Sector

The recreational fishing industry consists of enterprises such as the manufacture and retailing of fishing tackle and apparel, the payment of license fees to regulatory authorities, fishing books and magazines, the design and building of recreational fishing boats, and the provision of accommodation, fishing boats for charter, and guided fishing adventures.

## Aquaculture

Aquaculture, also known as aquafarming, is the farming of aquatic organisms such as fish, crustaceans, molluscs and aquatic plants. Aquaculture involves cultivating freshwater and saltwater populations under controlled conditions, and can be contrasted with commercial fishing, which is the harvesting of wild fish. Broadly speaking, finfish and shellfish fisheries can be conceptualized as akin to hunting and gathering while aquaculture is akin to agriculture. Mariculture refers to aquaculture practiced in marine environments and in underwater habitats.

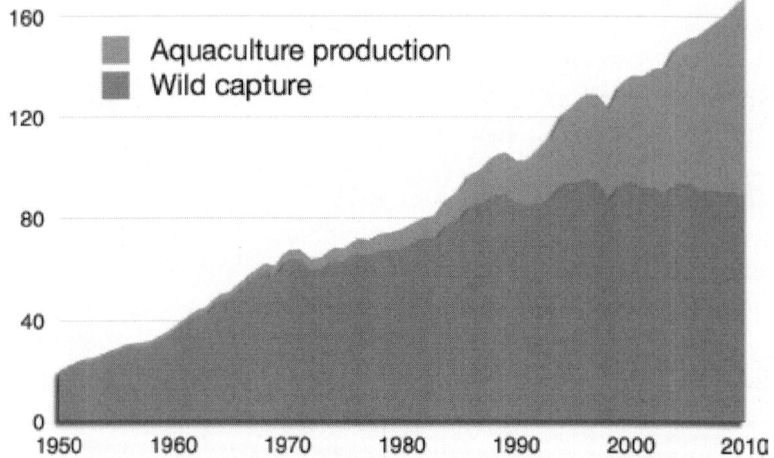


***Figure:*** *Global harvest of aquatic organisms in million tonnes, 1950–2010, as reported by the FAO*

According to the FAO, aquaculture "is understood to mean the farming of aquatic organisms including fish, molluscs, crustaceans and aquatic plants. Farming implies some form of intervention in the rearing process to enhance production, such as regular stocking, feeding, protection from predators, etc. Farming also implies individual or corporate ownership of the stock being cultivated." The reported output from global aquaculture operations would supply one half of the fish and shellfish that is directly consumed by humans; however, there are issues about the reliability of the reported figures. Further, in current aquaculture practice, products from several pounds of wild fish are used to produce one pound of a piscivorous fish like salmon.

Particular kinds of aquaculture include fish farming, shrimp farming, oyster farming, algaculture (such as seaweed farming), and the cultivation of ornamental fish. Particular methods include aquaponics and Integrated multi-trophic aquaculture, both of which integrate fish farming and plant farming.

The indigenous Gunditjmara people in Victoria, Australia may have raised eels as early as 6000 BC. There is evidence that they developed about 100 square kilometres (39 sq mi) of volcanic floodplains in the vicinity of Lake Condah into a complex of channels and dams, that they used woven traps to capture eels, and preserve eels to eat all year round.

Aquaculture was operating in China circa 2500 BC. When the waters subsided after river floods, some fishes, mainly carp, were trapped in lakes. Early aquaculturists fed their brood using nymphs and silkworm feces, and ate them. A fortunate genetic mutation of carp led to the emergence of goldfish during the Tang Dynasty.

Japanese cultivated seaweed by providing bamboo poles and, later, nets and oyster shells to serve as anchoring surfaces for spores.

### *Romans Bred Fish in Ponds*

In central Europe, early Christian monasteries adopted Roman aquacultural practices. Aquaculture spread in Europe during the Middle Ages since away from the seacoasts and the big rivers fish had to be salted in order to not perish. Improvements in transportation during the 19th century made fresh fish easily available and inexpensive, even in inland areas, making aquaculture less popular.

Hawaiians constructed oceanic fish ponds. A remarkable example is a fish pond dating from at least 1,000 years ago, at Alekoko. Legend says that it was constructed by the mythical Menehune dwarf people.

In 1859 Stephen Ainsworth of West Bloomfield, New York, began experiments with brook trout. By 1864 Seth Green had established a commercial fish hatching operation at Caledonia Springs, near Rochester, New York. By 1866, with the involvement of Dr. W. W. Fletcher of Concord, Massachusetts, artificial fish hatcheries were under way in both Canada and the United States. When the Dildo Island fish hatchery opened in Newfoundland in 1889, it was the largest and most advanced in the world.

Californians harvested wild kelp and attempted to manage supply circa 1900, later labeling it a wartime resource.

### *21st-Century Practice*

About 430 (97%) of the species cultured as of 2007 were domesticated during the 20th century, of which an estimated 106 came in the decade to 2007. Given the long-term importance of agriculture, it is interesting to note that to date only 0.08% of known land plant species and 0.0002% of known land animal species have been domesticated, compared with 0.17% of known marine plant species and 0.13% of known marine animal species. Domestication typically involves about a decade of scientific research. Domesticating aquatic species involves fewer risks to humans than land animals, which took a large toll in human lives. Most major human diseases originated in domesticated animals, through diseases such as smallpox and diphtheria, that like most infectious diseases, move to humans from animals. No human pathogens of comparable virulence have yet emerged from marine species.

Harvest stagnation in wild fisheries and overexploitation of popular marine species, combined with a growing demand for high quality protein, encourage aquaculturists to domesticate other marine species.

### *Species Groups*

Global aquaculture production in million tonnes, 1950–2010, as reported by the FAO

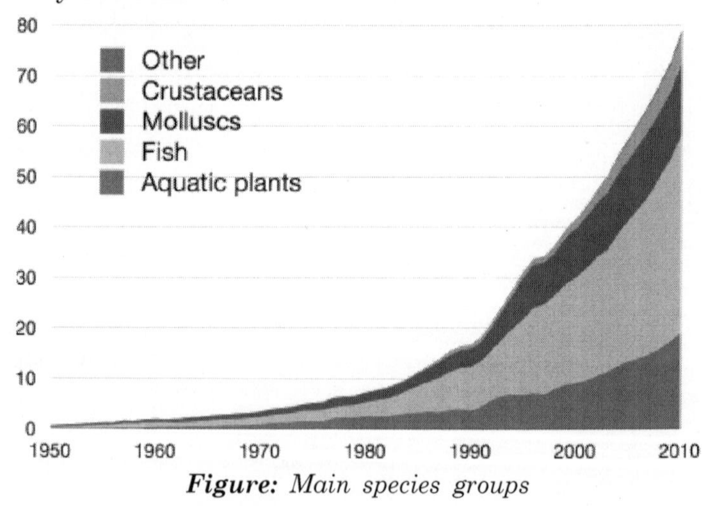


***Figure:*** *Main species groups*

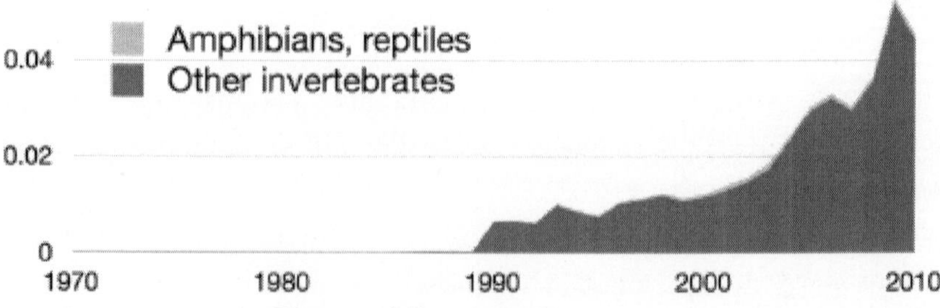


***Figure:*** *Minor species groups*

## Aquatic Plants

Microalgae, also referred to as phytoplankton, microphytes, or planktonic algae constitute the majority of cultivated algae. Macroalgae, commonly known as seaweed, also have many commercial and industrial uses, but due to their size and specific requirements, they are not easily cultivated on a large scale and are most often taken in the wild.

### *Fish*

The farming of fish is the most common form of aquaculture. It involves raising fish commercially in tanks, ponds, or ocean enclosures, usually for food. A facility that releases juvenile fish into the wild for recreational fishing or to supplement a species' natural numbers is generally referred to as a fish hatchery. Worldwide, the most important fish species used in fish farming are, in order, carp, salmon, tilapia and catfish.

In the Mediterranean, young bluefin tuna are netted at sea and towed slowly towards the shore. They are then interned in offshore pens where they are further grown for the market. In 2009, researchers in Australia managed for the first time to coax tuna (Southern bluefin) to breed in landlocked tanks.

### *Crustaceans*

Commercial shrimp farming began in the 1970s, and production grew steeply thereafter. Global production reached more than 1.6 million tonnes in 2003, worth about 9 billion U.S. dollars. About 75% of farmed shrimp is produced in Asia, in particular in China and Thailand. The other 25% is produced mainly in Latin America, where Brazil is the largest producer. Thailand is the largest exporter.

Shrimp farming has changed from its traditional, small-scale form in Southeast Asia into a global industry. Technological advances have led to ever higher densities per unit area, and broodstock is shipped worldwide. Virtually all farmed shrimp are penaeids (i.e., shrimp of the family *Penaeidae*), and just two species of shrimp, the Pacific white shrimp and the giant tiger prawn, account for about 80% of all farmed shrimp. These industrial monocultures are very susceptible to disease, which has decimated shrimp populations across entire regions. Increasing ecological problems, repeated disease outbreaks, and pressure and criticism from both NGOs and consumer countries led to changes in the industry in the late 1990s and generally stronger regulations. In 1999, governments, industry representatives, and environmental organisations initiated a program aimed at developing and promoting more sustainable farming practices through the Seafood Watch program.

Freshwater prawn farming shares many characteristics with, including many problems with marine shrimp farming. Unique problems are introduced by the developmental life cycle of the main species, the giant river prawn.

The global annual production of freshwater prawns (excluding crayfish and crabs) in 2003 was about 280,000 tonnes of which China produced 180,000 tonnes followed by India and Thailand with 35,000 tonnes each. Additionally, China produced about 370,000 tonnes of Chinese river crab.

### *Molluscs*

Aquacultured shellfish include various oyster, mussel and clam species. These bivalves are filter and/or deposit feeders, which rely on ambient primary production rather than inputs of fish or other feed. As such shellfish aquaculture is generally perceived as benign or even beneficial. Depending on the species and local conditions, bivalve molluscs are either grown on the beach, on longlines, or suspended from rafts and harvested by hand or by dredging. Abalone farming began in the late 1950s and early 1960s in Japan and China. Since the mid-1990s, this industry has become increasingly successful. Over-fishing and poaching have reduced wild populations to the extent that farmed abalone now supplies most abalone meat. Sustainably farmed molluscs can be certified by Seafood Watch and other organisations, including the World Wildlife Fund (WWF). WWF initiated the "Aquaculture Dialogues" in 2004 to develop measurable and performance-based standards for responsibly farmed seafood. In 2009, WWF co-founded the Aquaculture Stewardship Council (ASC) with the Dutch Sustainable Trade Initiative (IDH) to manage the global standards and certification programs.

### *Other Groups*

Other groups include aquatic reptiles, amphibians, and miscellaneous invertebrates, such as echinoderms and jellyfish. They are separately graphed at the top right of this section, since they do not contribute enough volume to show clearly on the main graph.

Commercially harvested echinoderms include sea cucumbers and sea urchins. In China, sea cucumbers are farmed in artificial ponds as large as 1,000 acres (400 ha).

### *Around the World*

Global aquaculture production in million tonnes, 1950–2010, as reported by the FAO

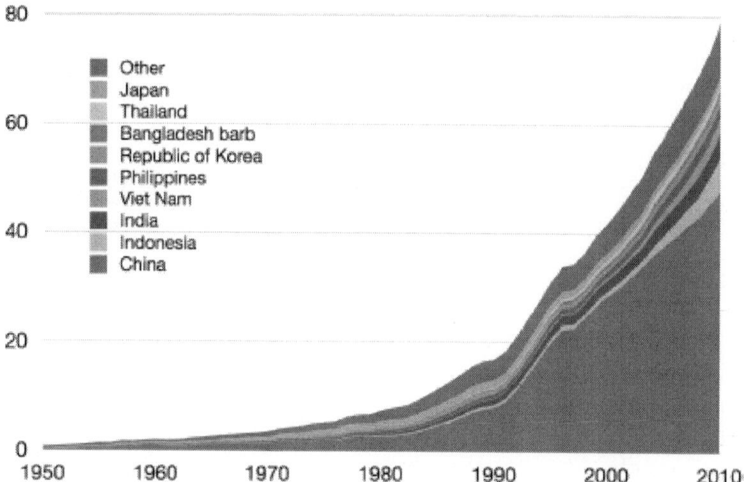


***Figure:*** *Main aquaculture countries, 1950–2010*

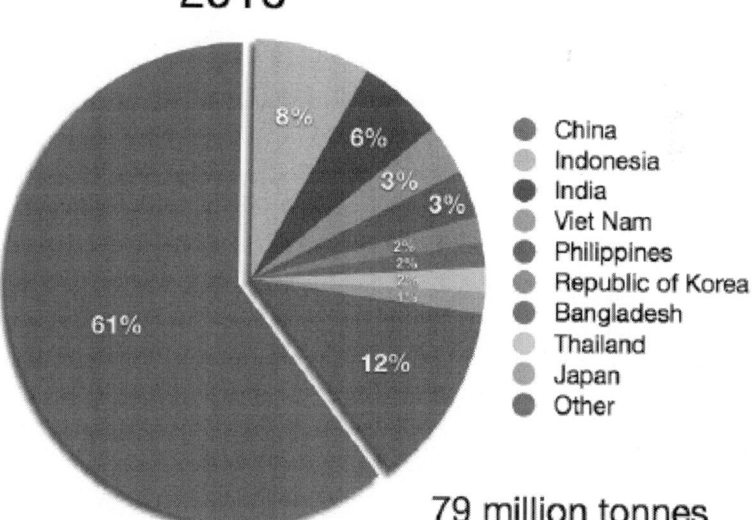


***Figure:*** *Main aquaculture countries in 2010*

In 2004, the total world production of fisheries was 140 million tonnes of which aquaculture contributed 45 million tonnes, about one third. The growth rate of worldwide aquaculture has been sustained and rapid, averaging about 8 percent per annum for over thirty years, while the take from wild fisheries has been essentially flat for the last decade. The aquaculture market reached $86 billion in 2009.

Aquaculture is an especially important economic activity in China. Between 1980 and 1997, the Chinese Bureau of Fisheries reports, aquaculture harvests grew at an annual rate of 16.7 percent, jumping from 1.9 million tonnes to nearly 23 million tonnes. In 2005, China accounted for 70% of world production. Aquaculture is also currently one of the fastest growing areas of food production in the U.S.

Approximately 90% of all U.S. shrimp consumption is farmed and imported. In recent years salmon aquaculture has become a major export in southern Chile, especially in Puerto Montt, Chile's fastest-growing city.

### Over Reporting

China overwhelmingly dominates the world in reported aquaculture output. They report a total output which is double that of the rest of the world put together. However, there are issues with the accuracy of China's returns.

In 2001, the fisheries scientists Reg Watson and Daniel Pauly expressed concerns in a letter to *Nature*, that China was over reporting its catch from wild fisheries in the 1990s. They said that made it appear that the global catch since 1988 was increasing annually by 300,000 tonnes, whereas it was really shrinking annually by 350,000 tonnes. Watson and Pauly suggested this may be related to China policies where state entities that monitor the economy are also tasked with increasing output. Also, until recently, the promotion of Chinese officials was based on production increases from their own areas.

China disputes this claim. The official Xinhua News Agency quoted Yang Jian, director general of the Agriculture Ministry's Bureau of Fisheries, as saying that China's figures were "basically correct". However, the FAO accepts there are issues with the reliability of China's statistical returns, and currently treats data from China, including the aquaculture data, apart from the rest of the world.

### Methods

***Mariculture:*** Mariculture is the term used for the cultivation of marine organisms in seawater, usually in sheltered coastal waters. In particular, the farming of marine fish is an example of mariculture, and so also is the farming of marine crustaceans (such as shrimps), molluscs (such as oysters) and seaweed.

***Figure:*** *Carp are the dominant fish in aquaculture*

***Figure:*** *The adaptable tilapia is another commonly farmed fish*

## Integrated

Integrated Multi-Trophic Aquaculture (IMTA) is a practice in which the by-products (wastes) from one species are recycled to become inputs (fertilizers, food) for another. Fed aquaculture (for example, fish, shrimp) is combined with inorganic extractive and organic extractive (for example, shellfish) aquaculture to create balanced systems for environmental sustainability (biomitigation), economic stability (product diversification and risk reduction) and social acceptability (better management practices).

"Multi-Trophic" refers to the incorporation of species from different trophic or nutritional levels in the same system. This is one potential distinction from the age-old practice of aquatic polyculture, which could simply be the co-culture of different fish species from the same trophic level. In this case, these organisms may all share the same biological and chemical processes, with few synergistic benefits, which could potentially lead to significant shifts in the ecosystem. Some traditional polyculture systems may, in fact, incorporate a greater diversity of species, occupying several niches, as extensive cultures (low intensity, low management) within the same pond. The "Integrated" in IMTA refers to the more intensive cultivation of the different species in proximity of each other, connected by nutrient and energy transfer through water.

Ideally, the biological and chemical processes in an IMTA system should balance. This is achieved through the appropriate selection and proportions of different species providing different ecosystem functions. The co-cultured species are typically more than just biofilters; they are harvestable crops of commercial value. A working IMTA system can result in greater total production based on mutual benefits to the co-cultured species and improved ecosystem health, even if the production of individual species is lower than in a monoculture over a short term period.

Sometimes the term "Integrated Aquaculture" is used to describe the integration of monocultures through water transfer. For all intents and purposes however, the terms "IMTA" and "integrated aquaculture" differ only in their degree of descriptiveness. Aquaponics, fractionated aquaculture, IAAS (integrated agriculture-aquaculture systems), IPUAS (integrated peri-urban-aquaculture systems), and IFAS (integrated fisheries-aquaculture systems) are other variations of the IMTA concept.

### *Netting Materials*

Various materials, including nylon, polyester, polypropylene, polyethylene, plastic-coated welded wire, rubber, patented rope products (Spectra, Thorn-D, Dyneema), galvanized steel and copper are used for netting in aquaculture fish enclosures around the world.

All of these materials are selected for a variety of reasons, including design feasibility, material strength, cost, and corrosion resistance.

Recently, copper alloys have become important netting materials in aquaculture because they are antimicrobial (i.e., they destroy bacteria, viruses, fungi, algae, and other microbes) and they therefore prevent biofouling (i.e., the undesirable accumulation, adhesion, and growth of microorganisms, plants, algae, tubeworms, barnacles, mollusks, and other organisms). By inhibiting microbial growth, copper alloy aquaculture cages avoid costly net changes that are necessary with other materials. The resistance of organism growth on copper alloy nets also provides a cleaner and healthier environment for farmed fish to grow and thrive.

### *Issues*

Aquaculture can be more environmentally damaging than exploiting wild fisheries on a local area basis but has considerably less impact on the global environment on a per kg of production basis. Local concerns include waste handling, side-effects of antibiotics, competition between farmed and wild animals, and using other fish to feed more marketable carnivorous fish. However, research and commercial feed improvements during the 1990s and 2000s have lessened many of these concerns.

Aquaculture may contribute to propagation of invasive species. As the cases of Nile perch and Janitor fish show, this issue may be damaging to native fauna.

Fish waste is organic and composed of nutrients necessary in all components of aquatic food webs. In-ocean aquaculture often produces much higher than normal fish waste concentrations. The waste collects on the ocean bottom, damaging or eliminating bottom-dwelling life. Waste can also decrease dissolved oxygen levels in the water column, putting further pressure on wild animals.

## Impacts on Wild Fish

Salmon farming currently leads to a high demand for wild forage fish. Fish do not actually produce omega-3 fatty acids, but instead accumulate them from either consuming microalgae that produce these fatty acids, as is the case with forage fish like herring and sardines, or, as is the case with fatty predatory fish, like salmon, by eating prey fish that have accumulated omega-3 fatty acids from microalgae. To satisfy this requirement, more than 50 percent of the world fish oil production is fed to farmed salmon.

In addition, as carnivores, salmon require large nutritional intakes of protein, protein which is often supplied to them in the form of forage fish. Consequently, farmed salmon consume more wild fish than they generate as a final product. To produce one pound of farmed salmon, products from several pounds of wild fish are fed to them. As the salmon farming industry expands, it requires more wild forage fish for feed, at a time when seventy five percent of the worlds monitored fisheries are already near to or have exceeded their maximum sustainable yield. The industrial scale extraction of wild forage fish for salmon farming then impacts the survivability of the wild predator fish who rely on them for food. Fish can escape from coastal pens, where they can interbreed with their wild counterparts, diluting wild genetic stocks. Escaped fish can become invasive, out competing native species.

### Coastal Ecosystems

Aquaculture is becoming a significant threat to coastal ecosystems. About 20 percent of mangrove forests have been destroyed since 1980, partly due to shrimp farming. An extended cost–benefit analysis of the total economic value of shrimp aquaculture built on mangrove ecosystems found that the external costs were much higher than the external benefits. Over four decades, 269,000 hectares (660,000 acres) of Indonesian mangroves have been converted to shrimp farms. Most of these farms are abandoned within a decade because of the toxin build-up and nutrient loss.

Salmon farms are typically sited in pristine coastal ecosystems which they then pollute. A farm with 200,000 salmon discharges more fecal waste than a city of 60,000 people. This waste is discharged directly into the surrounding aquatic environment, untreated, often containing antibiotics and pesticides." There is also an accumulation of heavy metals on the benthos (seafloor) near the salmon farms, particularly copper and zinc.

### Genetic Modification

A type of salmon have been genetically modified for faster growth, although it has not been approved for commercial use, due to opposition. One study, in a laboratory setting, found that modified salmon mixed with their wild relatives were aggressive in competing, but ultimately failed.

### Animal Welfare

As with the farming of terrestrial animals, social attitudes influence the need for humane practices and regulations in farmed marine animals. Under the guidelines advised by the Farm Animal Welfare Council good animal welfare means both fitness and a sense of well being in the animal's physical and mental state. This can be defined by the Five Freedoms:

- Freedom from hunger & thirst
- Freedom from discomfort
- Freedom from pain, disease, or injury
- Freedom to express normal behaviour
- Freedom from fear and distress

However, the controversial issue in aquaculture is whether fish and farmed marine invertebrates are actually sentient, or have the perception and awareness to experience suffering. Although no evidence of this has been found in marine invertebrates, recent studies conclude that fish do have the necessary receptors (nociceptors) to sense noxious stimuli and so are likely to experience states of pain, fear and stress. Consequently, welfare in aquaculture is directed at vertebrates; finfish in particular.

### *Common Welfare Concerns*

Welfare in aquaculture can be impacted by a number of issues such as stocking densities, behavioural interactions, disease and parasitism. A major problem in determining the cause of impaired welfare is that these issues are often all interrelated and influence each other at different times.

Optimal stocking density is often defined by the carrying capacity of the stocked environment and the amount of individual space needed by the fish, which is very species specific. Although behavioural interactions such as shoaling may mean that high stocking densities are beneficial to some species, in many cultured species high stocking densities may be of concern. Crowding can constrain normal swimming behaviour, as well as increase aggressive and competitive behaviours such as cannibalism, feed competition, territoriality and dominance/ subordination hierarchies. This potentially increases the risk of tissue damage due to abrasion from fish-to-fish contact or fish-to-cage contact. Fish can suffer reductions in food intake and food conversion efficiency. In addition, high stocking densities can result in water flow being insufficient, creating inadequate oxygen supply and waste product removal. Dissolved oxygen is essential for fish respiration and concentrations below critical levels can induce stress and even lead to asphyxiation. Ammonia, a nitrogen excretion product, is highly toxic to fish at accumulated levels, particularly when oxygen concentrations are low.

Many of these interactions and effects cause stress in the fish, which can be a major factor in facilitating fish disease. For many parasites, infestation depends on the host's degree of mobility, the density of the host population and vulnerability of the host's defence system. Sea lice are the primary parasitic problem for finfish in aquaculture, high numbers causing widespread skin erosion and haemorrhaging, gill congestion, and increased mucus production. There are also a number of prominent viral and bacterial pathogens that can have severe effects on internal organs and nervous systems.

### Improving Welfare

The key to improving welfare of marine cultured organisms is to reduce stress to a minimum, as prolonged or repeated stress can cause a range of adverse effects. Attempts to minimise stress can occur throughout the culture process. During grow out it is important to keep stocking densities at appropriate levels specific to each species, as well as separating size classes and grading to reduce aggressive behavioural interactions. Keeping nets and cages clean can assist positive water flow to reduce the risk of water degradation.

Not surprisingly disease and parasitism can have a major effect on fish welfare and it is important for farmers not only to manage infected stock but also to apply disease prevention measures. However, prevention methods, such as vaccination, can also induce stress because of the extra handling and injection. Other methods include adding antibiotics to feed, adding chemicals into water for treatment baths and biological control, such as using cleaner wrasse to remove lice from farmed salmon. Many steps are involved in transport, including capture, food deprivation to reduce faecal contamination of transport water, transfer to transport vehicle via nets or pumps, plus transport and transfer to the delivery location. During transport water needs to be maintained to a high quality, with regulated temperature, sufficient oxygen and minimal waste products. In some cases anaesthetics may be used in small doses to calm fish before transport.

Aquaculture is sometimes part of an environmental rehabilitation program or as an aid in conserving endangered species.

### Prospects

Global wild fisheries are in decline, with valuable habitat such as estuaries in critical condition. The aquaculture or farming of piscivorous fish, like salmon, does not help the problem because they need to eat products from other fish, such as fish meal and fish oil. Studies have shown that salmon farming has major negative impacts on wild salmon, as well as the forage fish that need to be caught to feed them. Fish that are higher on the food chain are less efficient sources of food energy.

Apart from fish and shrimp, some aquaculture undertakings, such as seaweed and filter-feeding bivalve mollusks like oysters, clams, mussels and scallops, are relatively benign and even environmentally restorative. Filter-feeders filter pollutants as well as nutrients from the water, improving water quality. Seaweeds extract nutrients such as inorganic nitrogen and phosphorus directly from the water, and filter-feeding mollusks can extract nutrients as they feed on particulates, such as phytoplankton and detritus.

Some profitable aquaculture cooperatives promote sustainable practices. New methods lessen the risk of biological and chemical pollution through minimizing fish stress, fallowing netpens, and applying Integrated Pest Management. Vaccines are being used more and more to reduce antibiotic use for disease control.

Onshore recirculating aquaculture systems, facilities using polyculture techniques, and properly sited facilities (for example, offshore areas with strong currents) are examples of ways to manage negative environmental effects.

Recirculating aquaculture systems (RAS) recycle water by circulating it through filters to remove fish waste and food and then recirculating it back into the tanks. This saves water and the waste gathered can be used in compost or, in some cases, could even be treated and used on land. While RAS was developed with freshwater fish in mind, scientist associated with the Agricultural Research Service have found a way to rear saltwater fish using RAS in low-salinity waters. Although saltwater fish are raised in offshore cages or caught with nets in water that typically has a salinity of 35 parts per thousand (ppt), scientists were able to produce healthy pompano, a saltwater fish, in tanks with a salinity of only 5 ppt. Commercializing low-salinity RAS are predicted to have positive environmental and economical effects. Unwanted nutrients from the fish food would not be added to the ocean and the risk of transmitting diseases between wild and farm-raised fish would greatly be reduced. The price of expensive saltwater fish, such as the pompano and combia used in the experiments, would be reduced. However, before any of this can be done researchers must study every aspect of the fish's lifecycle, including the amount of ammonia and nitrate the fish will tolerate in the water, what to feed the fish during each stage of its lifecycle, the stocking rate that will produce the healthiest fish, etc.

Some 16 countries now use geothermal energy for aquaculture, including China, Israel, and the United States. In California, for example, 15 fish farms produce tilapia, bass, and catfish with warm water from underground. This warmer water enables fish to grow all year round and mature more quickly. Collectively these California farms produce 4.5 million kilograms of fish each year.

## Recreational Fishing

Recreational fishing, also called sport fishing, is fishing for pleasure or competition. It can be contrasted with commercial fishing, which is fishing for profit, or subsistence fishing, which is fishing for survival.

The most common form of recreational fishing is done with a rod, reel, line, hooks and any one of a wide range of baits. Other devices, commonly referred to as *terminal tackle*, are also used to affect or complement the presentation of the bait to the targeted fish. Some examples of terminal tackle include weights, floats, and swivels. Lures are frequently used in place of bait. Some hobbyists make handmade tackle themselves, including plastic lures and artificial flies. The practice of catching or attempting to catch fish with a hook is known as angling.

Big-game fishing is conducted from boats to catch large open-water species such as tuna, sharks and marlin. Noodling and trout tickling are also recreational activities. One method of growing popularity is kayak fishing. Kayaks are stealthy and allow anglers to reach areas not fishable from land or by conventional boat. In addition, fishing from kayaks is regarded by some as an effort to level the playing field, to a degree, with their quarry and/or to challenge their angling abilities further by bringing an additional level of complexity to their sport. Historically, sport fishing has attracted greater interest among males. Women and girls represent barely 10% of the angling community, yet those who do enter the sport are often extremely successful, and at the highest levels of competitive angling, their results are comparable to those of their male counterparts.

Recreational fishing for sport or leisure gained popularity during the 16th and 17th centuries, and coincides with the publication of Izaak Walton's *The Compleat Angler, or Contemplative Man's*

*Recreation* in 1653. This book is the definitive work that champions the position of the angler who loves fishing for the sake of fishing.

More than 300 editions of *The Compleat Angler* have been published. The pastoral discourse is enriched with country fishing folklore, songs and poems, recipes and anecdotes, moral meditations, and quotes from classic literature. The central character, Piscator, champions the art of angling, but with an air of tranquility also relishes the pleasures of friendship, verse and song, and good food and drink.

The early evolution of fishing as recreation is not clear. For example, there is anecdotal evidence for fly fishing in Japan as early as the ninth century BCE, and in Europe Claudius Aelianus (175–235 CE) describes fly fishing in his work *On the Nature of Animals*, as

> *"a Macedonian way of catching fish... They fasten red (crimson red) wool round a hook, and fix on to the wool two feathers which grow under a cock's wattles, and which in colour are like wax. Their rod is six feet long, and their line is the same length. Then they throw their snare, and the fish, attracted and maddened by the colour, comes straight at it..."*

But for the early Japanese and Macedonians, fly fishing was likely to have been a means of survival, rather than recreation. It is possible that antecedents of recreational fly fishing arrived in England with the Norman conquest of 1066. Although the point in history where fishing could first be said to be recreational is not clear, it is clear that recreational fishing had fully arrived with the publication of *The Compleat Angler*.

Big-game fishing started as a sport after the invention of the motorized boat. In 1898, Dr. Charles Frederick Holder, a marine biologist and early conservationist, pioneered this sport and went on to publish many articles and books on the subject noted for their combination of accurate scientific detail with exciting narratives.

### Sport Fishing

Sport fishing methods vary according to the area fished, the species targeted, the personal strategies of the angler, and the resources available. It ranges from the aristocratic art of fly fishing elaborated in Great Britain, to the high-tech methods used to chase marlin and tuna. Sport fishing is usually done with hook, line, rod and reel rather than with nets or other aids. The most common salt water game fish are marlin, tuna, tarpon, sailfish, shark, and mackerel.

In North America, freshwater fish include snook, redfish, salmon, trout, bass, pike, catfish, walleye and muskellunge. The smallest fish are called panfish, because they can fit whole in a normal cooking pan. Examples are perch and sunfish.

In the past, sport fishers, even if they did not eat their catch, almost always killed them to bring them to shore to be weighed or for preservation as trophies. In order to protect recreational fisheries sport fishermen now often catch and release, and sometimes tag and release, which involves fitting the fish with identity tags, recording vital statistics, and sending a record to a government agency.

### *Fishing Techniques*

Recreational fishing techniques include hand gathering, spearfishing, netting, angling and trapping.

Most recreational fishers use a fishing rod with a fishing line and a hook at the end of the line. The rod may be equipped with a reel so the line can be reeled in, and some form of bait or a lure attached to the hook. Fly fishing is a special form of rod fishing in which the reel is attached to the back end of the rod, and heavy line is cast with a complex, repetitive whipping motion to deliver the ultralight artificial fly to its target. Another less common technique is bowfishing using a regular bow or a crossbow. The "arrow" is a modified bolt with barbs at the tip, connected to a fishing line so the fish can be retrieved. Some crossbows are fitted with a reel.

The effective use of fishing techniques often depends on knowledge about the fish and their behaviour including migration, foraging and habitat.

### *Fishing Tackle*

Fishing tackle is a general term that refers to the equipment used by fishers. Almost any equipment or gear used for fishing can be called fishing tackle. Some examples are hooks, lines, sinkers, floats, rods, reels, baits, lures, spears, nets, gaffs, traps, waders and tackle boxes.

Tackle that is attached to the end of a fishing line is called terminal tackle. This includes hooks, sinkers, floats, leaders, swivels, split rings and wire, snaps, beads, spoons, blades, spinners and clevises to attach spinner blades to fishing lures.

Fishing tackle can be contrasted with fishing techniques. Fishing tackle refers to the physical equipment that is used when fishing, whereas fishing techniques refers to the ways the tackle is used when fishing.

## Rules and Regulations

Recreational fishing has conventions, rules, licensing restrictions and laws that limit the way in which fish may be caught. The International Game Fish Association (IGFA) makes and oversees a set of voluntary guidelines. Typically, these prohibit the use of nets and the catching of fish with hooks not in the mouth. Enforceable regulations are put in place by governments to ensure sustainable practice amongst anglers. For example in the Republic of Ireland, the Central Fisheries Board oversees the implementation of all angling regulations, which include controls on angling lures, baits and number of hooks permissible, as well as licensing requirements and other conservation-based restrictions. Regulations notwithstanding, voluntary catch and release fishing as a means of protecting and sustaining game species has become an increasingly common practice among conservation-minded recreational anglers.

## Fish Logs

In addition to capturing fish for food, recreational anglers might also keep a log of fish caught and submit trophy-sized fish to independent record keeping bodies. In the Republic of Ireland, the Irish Specimen Fish Committee verifies and publicizes the capture of trophy fish caught with rod and line by anglers in Ireland, both in freshwater and at sea. The Committee also ratifies Irish record rod caught fish. It also uses a set of 'fair play' regulations to ensure fish are caught in accordance with accepted angling norms.

## Competitions

Recreational fishing competitions (tournaments) are a recent innovation in which fishermen compete for prizes based on the total weight of a given species of fish caught within a predetermined time. This sport evolved from local fishing contests into large competitive circuits, especially in North America. Competitors are most often professional fishermen who are supported by commercial endorsements. Other competitions are based purely on length with mandatory catch and release. Either longest fish or total length is documented with a camera and a mandatory sticker or unique item, a practice used since it's hard to weigh a living fish accurately in a boat.

Sport fishing competitions involve individuals if the fishing occurs from land, and usually teams if conducted from boats, as well as specified times and areas for catching fish. A score is awarded for each fish caught. The points awarded depend on the fish's weight and species. Occasionally a score is divided by the strength of the fishing line used, yielding more points to those who use thinner, weaker line. In tag and release competitions, a flat score is awarded per fish species caught, divided by the line strength. Usually sport fishing competitions award a prize to the boat or team with the most points earned.

In Australia, a self-administered standard for the environmental assessment of tournament fishing has been proposed as an alternative and possible pathway to the ISO 14001 international standard. The standard assesses environmental, social, economic, and public risk factors. Tournament organisers may apply for voluntary certification. In some US states, fishery agencies and competition organisers create their own codes of practice.

### *Industry*

The recreational fishing industry consists of enterprises such as the manufacture and retailing of fishing tackle, the design and building of recreational fishing boats, and the provision of fishing boats for charter and guided fishing trips.

"Pay to fish" enterprises provide anglers with controlled access to stocked lakes, ponds, or canals. These provide fishing opportunities outside of the permitted seasons and quotas applied to public waters. In the United Kingdom, commercial fisheries of this sort charge access fees. In North America, establishments usually charge for the fish caught, by length or by weight, rather than for access to the site although some establishments charge both types of fees. Recreational fishing is a multi-billion dollar industry In the USA, about 12 million recreational saltwater fishers generate $30 billion in economic impact and support 350,000 jobs.

## Fishing (Fishing for Food)

Fishing is the activity of trying to catch fish. Fish are normally caught in the wild. Techniques for catching fish include hand gathering, spearing, netting, angling and trapping. The term fishing may be applied to catching other aquatic animals such as molluscs, cephalopods, crustaceans, and echinoderms. The term is not normally applied to catching farmed fish, or to aquatic mammals, such as whales, where the term whaling is more appropriate.

According to FAO statistics, the total number of commercial fishermen and fish farmers is estimated to be 38 million. Fisheries and aquaculture provide direct and indirect employment to over 500 million people. In 2005, the worldwide per capita consumption of fish captured from wild fisheries was 14.4 kilograms, with an additional 7.4 kilograms harvested from fish farms. In addition to providing food, modern fishing is also a recreational pastime.

## History

Fishing is an ancient practice that dates back to at least the beginning of the Paleolithic period about 40,000 years ago. Isotopic analysis of the skeletal remains of Tianyuan man, a 40,000-year old modern human from eastern Asia, has shown that he regularly consumed freshwater fish. Archaeology features such as shell middens, discarded fish bones and cave paintings show that sea foods were important for survival and consumed in significant quantities. During this period, most people lived a hunter-gatherer lifestyle and were, of necessity, constantly on the move. However, where there are early examples of permanent settlements (though not necessarily permanently occupied) such as those at Lepenski Vir, they are almost always associated with fishing as a major source of food.

The ancient river Nile was full of fish; fresh and dried fish were a staple food for much of the population. The Egyptians had implements and methods for fishing and these are illustrated in tomb scenes, drawings, and papyrus documents. Some representations hint at fishing being pursued as a pastime. In India, the Pandyas, a classical Dravidian Tamil kingdom, were known for the pearl fishery as early as the 1st century BC. Their seaport Tuticorin was known for deep sea pearl fishing.

The Paravas, a Tamil caste centred in Tuticorin, developed a rich community because of their pearl trade, navigation knowledge and fisheries. Seafood played a central role in the food culture of ancient Greeks, though fishing scenes are rarely represented in their art, a reflection of the low social status of fishing. Oppian of Corycus, a Greek author wrote a major treatise on sea fishing; the *Halieulica* or *Halieutika*, composed between 177 and 180. This is the earliest such work to have survived to the modern day. Pictorial evidence of Roman fishing comes from mosaics. The Roman god of the sea Neptune is depicted as wielding a fishing trident. The Moche people of ancient Peru depicted fishermen in their ceramics.

One of the world's longest trading histories is the trade of dry cod from the Lofoten area of Norway to the southern parts of Europe, Italy, Spain and Portugal. The trade in cod started during the Viking period or before, has been going on for more than 1,000 years and is still important.

### Techniques

There are many fishing techniques and tactics for catching fish. The term can also be applied to methods for catching other aquatic animals such as molluscs (shellfish, squid, octopus) and edible marine invertebrates.

Fishing techniques include hand gathering, spearfishing, netting, angling and trapping. Recreational, commercial and artisanal fishers use different techniques, and also, sometimes, the same techniques. Recreational fishers fish for pleasure or sport, while commercial fishers fish for profit. Artisanal fishers use traditional, low-tech methods, for survival in third-world countries, and as a cultural heritage in other countries. Mostly, recreational fishers use angling methods and commercial fishers use netting methods. There is an intricate link between various fishing techniques and knowledge about the fish and their behaviour including migration, foraging and habitat. The effective use of fishing techniques often depends on this additional knowledge. Some fishermen follow fishing folklores which claim that fish feeding patterns are influenced by the position of the sun and the moon.

### Tackle

Fishing tackle is a general term that refers to the equipment used by fishermen when fishing. Almost any equipment or gear used for fishing can be called fishing tackle. Some examples are hooks, lines, sinkers, floats, rods, reels, baits, lures, spears, nets, gaffs, traps, waders and tackle boxes.

Tackle that is attached to the end of a fishing line is called terminal tackle. This includes hooks, sinkers, floats, leaders, swivels, split rings and wire, snaps, beads, spoons, blades, spinners and clevises to attach spinner blades to fishing lures. Fishing tackle can be contrasted with fishing techniques. Fishing tackle refers to the physical equipment that is used when fishing, whereas fishing techniques refers to the ways the tackle is used when fishing.

### Fishing Vessels

A fishing vessel is a boat or ship used to catch fish in the sea, or on a lake or river. Many different kinds of vessels are used in commercial, artisanal and recreational fishing.

According to the FAO, in 2004 there were four million commercial fishing vessels. About 1.3 million of these are decked vessels with enclosed areas. Nearly all of these decked vessels are mechanised, and 40,000 of them are over 100 tons. At the other extreme, two-thirds (1.8

million) of the undecked boats are traditional craft of various types, powered only by sail and oars. These boats are used by artisan fishers.

It is difficult to estimate how many recreational fishing boats there are, although the number is high. The term is fluid, since most recreational boats are also used for fishing from time to time. Unlike most commercial fishing vessels, recreational fishing boats are often not dedicated just to fishing. Just about anything that will stay afloat can be called a recreational fishing boat, so long as a fisher periodically climbs aboard with the intent to catch a fish.

Fish are caught for recreational purposes from boats which range from dugout canoes, kayaks, rafts, pontoon boats and small dingies to runabouts, cabin cruisers and cruising yachts to large, hi-tech and luxurious big game rigs. Larger boats, purpose-built with recreational fishing in mind, usually have large, open cockpits at the stern, designed for convenient fishing.

### Traditional Fishing

Traditional fishing is any kind of small scale, commercial or subsistence fishing practices using traditional techniques such as rod and tackle, arrows and harpoons, throw nets and drag nets, etc.

### Fish Farms

Fish farming is the principal form of aquaculture, while other methods may fall under mariculture. It involves raising fish commercially in tanks or enclosures, usually for food. A facility that releases juvenile fish into the wild for recreational fishing or to supplement a species' natural numbers is generally referred to as a fish hatchery. Fish species raised by fish farms include salmon, carp, tilapia, catfish, trout and others.

Increased demands on wild fisheries by commercial fishing has caused widespread overfishing. Fish farming offers an alternative solution to the increasing market demand for fish and fish protein.

## Fisheries Management

Fisheries management draws on fisheries science in order to find ways to protect fishery resources so sustainable exploitation is possible. Modern fisheries management is often referred to as a governmental system of (hopefully appropriate) management rules based on defined objectives and a mix of management means to implement the rules, which are put in place by a system of monitoring control and surveillance.

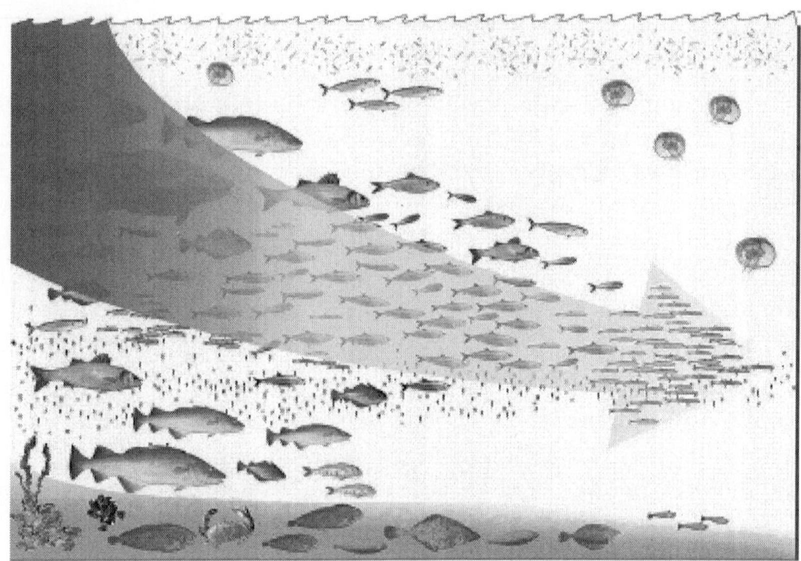

***Figure:*** *Fishing down the food web*

Fisheries science is the academic discipline of managing and understanding fisheries. It is a multidisciplinary science, which draws on the disciplines of oceanography, marine biology, marine conservation, ecology, population dynamics, economics and management in an attempt to provide an integrated picture of fisheries. In some cases new disciplines have emerged, such as bioeconomics.

### Sustainability

Issues involved in the long term sustainability of fishing include overfishing, by-catch, marine pollution, environmental effects of fishing, climate change and fish farming.

Conservation issues are part of marine conservation, and are addressed in fisheries science programs. There is a growing gap between how many fish are available to be caught and humanity's desire to catch them, a problem that gets worse as the world population grows.

Similar to other environmental issues, there can be conflict between the fishermen who depend on fishing for their livelihoods and fishery scientists who realise that if future fish populations are to be sustainable then some fisheries must limit fishing or cease operations.

### Cultural Impact

- Community impact: For communities like fishing villages, fisheries provide not only a source of food and work but also a community and cultural identity.

- Semantic impact: A "fishing expedition" is a situation where an interviewer implies he knows more than he actually does in order to trick his target into divulging more information than he wishes to reveal. Other examples of fishing terms that carry a negative connotation are: "fishing for compliments", "to be fooled hook, line and sinker" (to be fooled beyond merely "taking the bait"), and the internet scam of Phishing in which a third party will duplicate a website where the user would put sensitive information (such as bank codes).
- Religious impact: Fishing has had an effect on all major religions, including Islam, Christianity, Buddhism, Jainism, Zoroastrianism, Hinduism, and the various new age religions. Jesus was known to participate in fishing excursions. According to the Roman Catholic faith the first Pope was a fisherman, the Apostle Peter, a number of the miracles, and many parables and stories reported in the Bible involve it. The Pope's traditional vestments include a fish-shaped hat.

# 4

# Seafood

Seafood is any form of sea life regarded as food by humans. Seafood prominently includes fish and shellfish. Shellfish include various species of molluscs, crustaceans, and echinoderms. Historically, sea mammals such as whales and dolphins have been consumed as food, though that happens to a lesser extent these days. Edible sea plants, such as some seaweeds and microalgae, are widely eaten as seafood around the world, especially in Asia. In North America, although not generally in the United Kingdom, the term "seafood" is extended to fresh water organisms eaten by humans, so all edible aquatic life may be referred to as seafood. For the sake of completeness, this article includes all edible aquatic life.

The harvesting of wild seafood is known as fishing and the cultivation and farming of seafood is known as aquaculture, mariculture, or in the case of fish, fish farming. Seafood is often distinguished from meat, although it is still animal and is excluded in a strict vegetarian diet. Seafood is an important source of protein in many diets around the world, especially in coastal areas.

Most of the seafood harvest is consumed by humans, but a significant proportion is used as fish food to farm other fish or rear farm animal. Some seafoods (kelp) are used as food for other plants (fertilizer). In these ways, seafoods are indirectly used to produce further food for human consumption. Products, such as fish oil and spirulina tablets are also extracted from seafoods. Some seafood is feed to aquarium fish, or used to feed domestic pets, such as cats, and a small proportion is used in medicine, or is used industrially for non-food purposes (leather).

## History

The harvesting, processing, and consuming of seafoods are ancient practices that date back to at least the beginning of the Paleolithic period about 40,000 years ago. Isotopic analysis of the skeletal remains of Tianyuan man, a 40,000 year old modern human from eastern Asia, has shown that he regularly consumed freshwater fish. Archaeology features such as shell middens, discarded fish bones and cave paintings show that sea foods were important for survival and consumed in significant quantities. During this period, most people lived a hunter-gatherer lifestyle and were, of necessity, constantly on the move. However, where there are early examples of permanent settlements (though not necessarily permanently occupied) such as those at Lepenski Vir, they are almost always associated with fishing as a major source of food.

The ancient river Nile was full of fish; fresh and dried fish were a staple food for much of the population. The Egyptians had implements and methods for fishing and these are illustrated in tomb scenes, drawings, and papyrus documents. Some representations hint at fishing being pursued as a pastime.

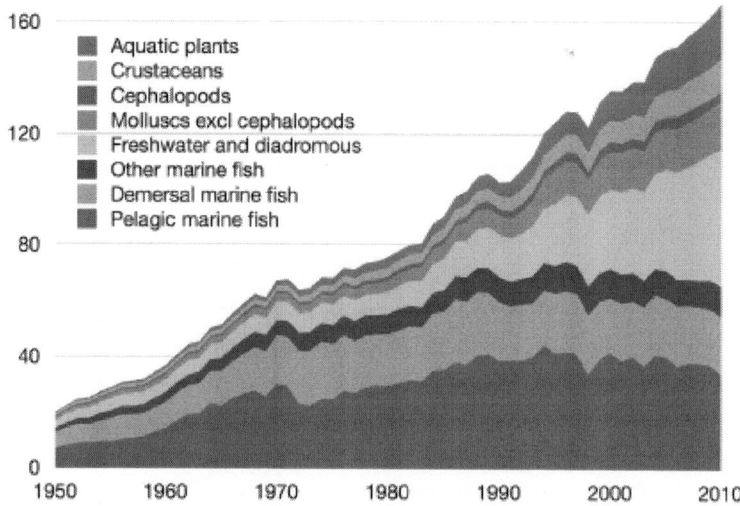


***Figure:*** *World fisheries harvest, both wild and farmed, in million tonnes, 1950–2010*

Fishing scenes are rarely represented in ancient Greek culture, a reflection of the low social status of fishing. However, Oppian of Corycus, a Greek author wrote a major treatise on sea fishing, the *Halieulica* or *Halieutika*, composed between 177 and 180. This is the earliest such work to have survived to the modern day. The consumption of fish varied in accordance with the wealth and location of the household. In the Greek islands and on the coast, fresh fish and seafood (squid, octopus, and shellfish) were common. They were eaten locally but more often transported inland. Sardines and anchovies were regular fare for the citizens of Athens. They were sometimes sold fresh, but more frequently salted. A stele of the late 3rd century BCE from the small Boeotian city of Akraiphia, on Lake Copais, provides us with a list of fish prices.

The cheapest was *skaren* (probably parrotfish) whereas Atlantic bluefin tuna was three times as expensive. Common salt water fish were yellowfin tuna, red mullet, ray, swordfish or sturgeon, a delicacy which was eaten salted. Lake Copais itself was famous in all Greece for its eels, celebrated by the hero of *The Acharnians*. Other fresh water fish were pike-fish, carp and the less appreciated catfish.

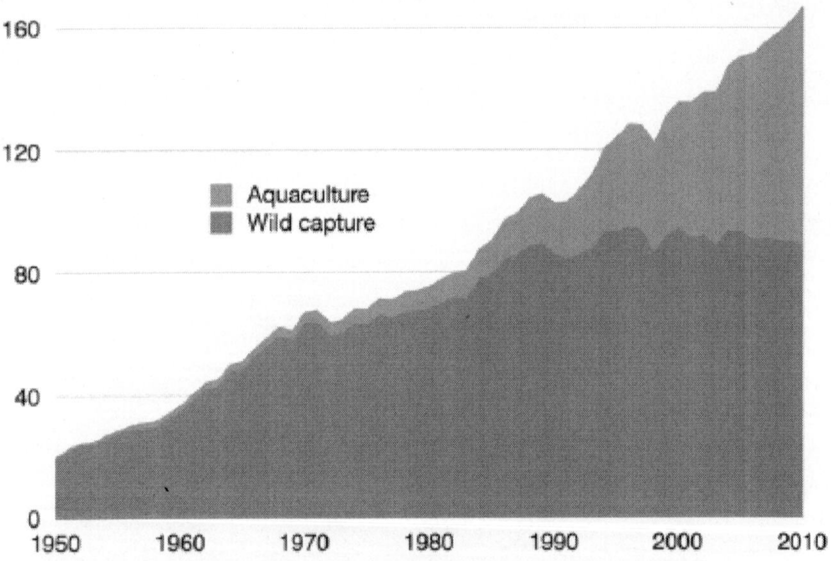


***Figure:*** *World fisheries harvest, wild capture versus aquaculture production, in million tonnes 1950–2010*

Pictorial evidence of Roman fishing comes from mosaics. At a certain time the goatfish was considered the epitome of luxury, above all because its scales exhibit a bright red colour when it dies out of water. For this reason these fish were occasionally allowed to die slowly at the table. There even was a recipe where this would take place *in garo*, in the sauce. At the beginning of the Imperial era, however, this custom suddenly came to an end, which is why *mullus* in the feast of Trimalchio could be shown as a characteristic of the *parvenu*, who bores his guests with an unfashionable display of dying fish.

In medieval times, seafood was less prestigious than other animal meats, and often seen as merely an alternative to meat on fast days. Still, seafood was the mainstay of many coastal populations. Kippers made from herring caught in the North Sea could be found in markets as far away as Constantinople. While large quantities of fish were eaten fresh, a large proportion was salted, dried, and, to a lesser extent, smoked. Stockfish, cod that was split down the middle, fixed to a pole and dried, was very common, though preparation could be time-consuming, and meant beating the dried fish with a mallet before soaking it in water. A wide range of mollusks including oysters, mussels and scallops were eaten by coastal and river-dwelling populations, and freshwater crayfish were seen as a desirable alternative to meat during fish days. Compared to meat, fish was much more expensive for inland populations, especially in Central Europe, and therefore not an option for most.

Modern knowledge of the reproductive cycles of aquatic species has led to the development of hatcheries and improved techniques of fish farming and aquaculture. Better understanding of the hazards of eating raw and undercooked fish and shellfish has led to improved preservation methods and processing.

### Processing

Fish is a highly perishable product. The *fishy* smell of dead fish is due to the breakdown of amino acids into biogenic amines and ammonia. Live food fish are often transported in tanks at high expense for an international market that prefers its seafood killed immediately before it is cooked. This process originally was started by Lindeye. Delivery of live fish without water is also being explored. While some seafood restaurants keep live fish in aquaria for display purposes or for cultural beliefs, the majority of live fish are kept for dining customers. The live food fish trade in Hong Kong, for example, is estimated to have driven imports of live food fish to more than 15,000 tonnes in 2000. Worldwide sales that year were estimated at US$400 million, according to the World Resources Institute.

If the cool chain has not been adhered to correctly, food products generally decay and become harmful before the validity date printed on the package. As the potential harm for a consumer when eating rotten fish is much larger than for example with dairy products, the

U.S. Food and Drug Administration (FDA) has introduced regulation in the USA requiring the use of a time temperature indicator on certain fresh chilled seafood products.

Fresh fish is a highly perishable food product, so it must be eaten promptly or discarded; it can be kept for only a short time. In many countries, fresh fish are filleted and displayed for sale on a bed of crushed ice or refrigerated. Fresh fish is most commonly found near bodies of water, but the advent of refrigerated train and truck transportation has made fresh fish more widely available inland.

Long term preservation of fish is accomplished in a variety of ways. The oldest and still most widely used techniques are drying and salting. Desiccation (complete drying) is commonly used to preserve fish such as cod. Partial drying and salting is popular for the preservation of fish like herring and mackerel. Fish such as salmon, tuna, and herring are cooked and canned. Most fish are filleted prior to canning, but some small fish (e.g. sardines) are only decapitated and gutted prior to canning.

### Mislabeling

As the most traded food commodity worldwide, seafood has to be labeled correctly in order to protect our oceans, consumers and the economy. However, due to globalization, fraud is becoming increasingly common in seafood trade.

A recent study by the organisation Oceana revealed that over one-third of collected seafood samples in the U.S. has not been labeled correctly. Snapper and tuna occupy a sad first place with the highest rates of mislabeled products. Seafood substitution is the best known type of seafood fraud, according to the Oceana report 2013. This doesn't only harm the consumers' wallet but also poses health risks. Another type of mislabeling is the so-called short-weighting. Practices such as overglazing or soaking increase the net weight of the fish. The consumers pay more for less in the end. The detection of water retention agents helps identify the fraud and its origin.

### Consumption

Seafood is consumed all over the world; it provides the world's prime source of high-quality protein: 14–16% of the animal protein consumed world-wide; over one billion people rely on seafood as their primary source of animal protein. Fish is among the most common food allergens.

Iceland, Japan, and Portugal are the greatest consumers of seafood per capita in the world.

The UK Food Standards Agency recommends that at least two portions of seafood should be consumed each week, one of which should be oil-rich. There are over 100 different types of seafood available around the coast of the UK.

Oil-rich fish such as mackerel or herring are rich in long chain Omega-3 oils. These oils are found in every cell of the human body, and are required for human biological functions such as brain functionality.

Whitefish such as haddock and cod are very low in fat and calories which, combined with oily fish rich in Omega-3 such as mackerel, sardines, fresh tuna, salmon and trout, can help to protect against coronary heart disease, as well as helping to develop strong bones and teeth.

Shellfish are particularly rich in zinc, which is essential for healthy skin and muscles as well as fertility. Casanova reputedly ate 50 oysters a day.

### Health Benefits

Research over the past few decades has shown that the nutrients and minerals in seafood can make improvements in brain development and reproduction and has highlighted the role for seafood in the functions of the human body.

Doctors have known of strong links between fish and healthy hearts ever since they noticed that fish-eating Inuit populations in the Arctic had low levels of heart disease. One study has suggested that adding one portion of fish a week to your diet can cut your chances of suffering a heart attack by half. Fish is thought to protect the heart because eating less saturated fat and more Omega-3 can help to lower the amount of cholesterol and triglycerides in the blood – two fats that, in excess, increase the risk of heart disease. Omega-3 fats also have natural built-in anti-oxidants, which are thought to stop the thickening and damaging of artery walls. Regularly eating fish oils is also thought to reduce the risk of arrhythmia – irregular electrical activity in the heart which increases the risk of sudden heart attacks.

10-12% of the human brain is composed of lipids, including the Omega-3 fat DHA. Recent studies suggest that older people can boost their brain power by eating more oily fish, what with regular consumers being able to remember better and think faster than those who don't consume at all. Other research has also suggested that adding more DHA to the diet of children with attention-deficit hyperactivity disorder can reduce their behavioural problems and improve their reading skills, while there have also been links suggested between DHA and better concentration. Separate studies have suggested that older people who eat fish at least once a week could also have a lower chance of developing dementia and Alzheimer's disease.

Including fish as a regular part of a balanced diet has been shown to help the symptoms of rheumatoid arthritis – a painful condition that causes joints to swell up, reducing strength and mobility. Studies also show that sufferers feel less stiff and sore in the morning if they keep their fish oil intake topped up. Recent research has also found a link between Omega-3 fats and a slowing down in the wearing of cartilage that leads to osteoarthritis, opening the door for more research into whether eating more fish could help prevent the disease.

Fish is high in minerals such as zinc, iodine and selenium, which keep the body running smoothly. Iodine is essential for the thyroid gland, which controls growth and metabolism, while selenium is used to make enzymes that protect cell walls from cancer-causing free radicals, and helps prevent DNA damage caused by radiation and some chemicals. Fish is also a source of vitamin A, which is needed for healthy skin and eyes, and vitamin D, which is needed to help the body absorb calcium to strengthen teeth and bones.

### Health Hazards

Fish and shellfish have a natural tendency to concentrate mercury in their bodies, often in the form of methylmercury, a highly toxic organic compound of mercury. Species of fish that are high on the food chain, such as shark, swordfish, king mackerel, albacore tuna, and tilefish contain higher concentrations of mercury than others. This is because mercury is stored in the muscle tissues of fish, and when a predatory fish eats another fish, it assumes the entire body burden of mercury in the consumed fish. Since fish are less efficient at depurating than accumulating methylmercury, fish-tissue concentrations increase over time. Thus species that are high on the food chain amass body burdens of mercury that can be ten times higher than the species they consume. This process is called biomagnification. The first occurrence of widespread mercury poisoning in humans occurred this way in Minamata, Japan, now called Minamata disease.

Shellfish are among the more common food allergens.

### Sustainability

Research into population trends of various species of seafood is pointing to a global collapse of seafood species by 2048. Such a collapse would occur due to pollution and overfishing, threatening oceanic ecosystems, according to some researchers.

A major international scientific study released in November 2006 in the journal *Science* found that about one-third of all fishing stocks worldwide have collapsed (with a collapse being defined as a decline to less than 10% of their maximum observed abundance), and that if current trends continue all fish stocks worldwide will collapse within fifty years. In July 2009, Boris Worm of Dalhousie University, the author of the November 2006 study in *Science,* co-authored an update on the state of the world's fisheries with one of the original study's critics, Ray Hilborn of the University of Washington at Seattle. The new study found that through good fisheries management techniques even depleted fish stocks can be revived and made commercially viable again.

The FAO State of World Fisheries and Aquaculture 2004 report estimates that in 2003, of the main fish stocks or groups of resources for which assessment information is available, "approximately one-quarter were overexploited, depleted or recovering from depletion (16%, 7% and 1% respectively) and needed rebuilding."

The National Fisheries Institute, a trade advocacy group representing the United States seafood industry, disagree. They claim that currently observed declines in fish population are due to natural fluctuations and that enhanced technologies will eventually alleviate whatever impact humanity is having on oceanic life.

### In Religion

For the most part Islamic dietary laws allow the eating of seafood, though the Hanbali forbid eels, the Shafi forbid frogs and crocodiles, and the Hanafi forbid bottom feeders such as shellfish and carp. The Jewish laws of Kashrut forbid the eating of shellfish and eels. According to the King James version of the bible, it is alright to eat finfish, but shellfish and eels are an abomination and should not be eaten. Since early times, the Catholic church has forbidden the practice of eating meat, eggs and dairy products at certain times. Thomas Aquinas argued that these "afford greater pleasure as food [than fish], and greater nourishment to the human body, so that from their consumption there results a greater surplus available for seminal matter, which when abundant becomes a great incentive to lust." In the United

States, the Catholic practice of abstaining from meat on Fridays has popularized the Friday fish fry, and parishes often sponsor a fish fry during Lent. In predominantly Roman Catholic areas, restaurants may adjust their menus during Lent by adding seafood items to the menu.

## Biology of Deep Sea-Fish

One of the concerns with inshore aquaculture is that discarded nutrients and feces can settle below the farm on the seafloor and damage the benthic ecosystem. According to its proponents, the wastes from aquaculture that has been moved offshore tend to be swept away from the site and diluted. Moving aquaculture offshore also provides more space where aquaculture production can expand to meet the increasing demands for fish. It avoids many of the conflicts that occur with other marine resource users in the more crowded inshore waters, though there can still be user conflicts offshore. Critics are concerned about issues such as the ongoing consequences of using antibiotics and other drugs and the possibilities of cultured fish escaping and spreading disease among wild fish.

Deep-sea fish are fish that live in the darkness below the sunlit surface waters, that is below the epipelagic or photic zone of the ocean. The lanternfish is, by far, the most common deep-sea fish. Other deep sea fish include the flashlight fish, cookiecutter shark, bristlemouths, anglerfish, and viperfish.

Only about 2% of known marine species inhabit the pelagic environment. This means that they live in the water column as opposed to the benthic organisms that live in or on the sea floor. Deep-sea organisms generally inhabit bathypelagic (1000m-4000m deep) and abyssopelagic (4000m-6000m deep) zones. However, characteristics of deep-sea organisms, such as bioluminescence can be seen in the mesopelagic (200m-1000m deep) zone as well. The mesopelagic zone is the disphotic zone, meaning light there is minimal but still measurable. The oxygen minimum layer exists somewhere between a depth of 700m and 1000m deep depending on the place in the ocean. This area is also where nutrients are most abundant. The bathypelagic and abyssopelagic zones are aphotic, meaning that no light penetrates this area of the ocean. These zones make up about 75% of the inhabitable ocean space.

The zone that deep-sea fish do not inhabit is the epipelagic zone (0m-200m), which is the area where light penetrates the water and photosynthesis occurs. This is also known as the euphotic, or more simply as the photic zone. Because the photic zone typically extends only a few hundred metres below the water, about 90% of the ocean volume is in darkness. The deep-sea is also an extremely hostile environment, with temperatures that rarely exceed 3°C and fall as low as -1.8°C" (with the exception of hydrothermal vent ecosystems that can exceed 350°C), low oxygen levels, and pressures between 20 and 1,000 atmospheres (between 2 and 100 megapascals).

### *Environment*

In the deep ocean, the waters extend far below the epipelagic zone, and support very different types of pelagic fishes adapted to living in these deeper zones.

In deep water, marine snow is a continuous shower of mostly organic detritus falling from the upper layers of the water column. Its origin lies in activities within the productive photic zone. Marine snow includes dead or dying plankton, protists (diatoms), fecal matter, sand, soot and other inorganic dust. The "snowflakes" grow over time and may reach several centimetres in diameter, travelling for weeks before reaching the ocean floor. However, most organic components of marine snow are consumed by microbes, zooplankton and other filter-feeding animals within the first 1,000 metres of their journey, that is, within the epipelagic zone. In this way marine snow may be considered the foundation of deep-sea mesopelagic and benthic ecosystems: As sunlight cannot reach them, deep-sea organisms rely heavily on marine snow as an energy source.

Some deep-sea pelagic groups, such as the lanternfish, ridgehead, marine hatchetfish, and lightfish families are sometimes termed *pseudoceanic* because, rather than having an even distribution in open water, they occur in significantly higher abundances around structural oases, notably seamounts and over continental slopes. The phenomenon is explained by the likewise abundance of prey species which are also attracted to the structures.

Hydrostatic pressure increases by 1 atmosphere for every 10m in depth. Deep-sea organisms have the same pressure within their bodies that is being exerted on them from the outside, so they aren't crushed by the extreme pressure. Their high internal pressure, however, results in the reduced fluidity of their membranes because molecules are squeezed together. Fluidity in cell membranes increases efficiency of biological functions, most importantly the production of proteins, so organisms have adapted to this circumstance by increasing the proportion of unsaturated fatty acids in the lipids of the cell membranes. In addition to differences in internal pressure, these organisms have developed a different balance between their metabolic reactions from those organisms that live in the epipelagic zone. David Wharton, author of Life at the Limits: Organisms in Extreme Environments notes, "Biochemical reactions are accompanied by changes in volume. If a reaction results in an increase in volume, it will be inhibited by pressure, whereas, if it is associated with a decrease in volume, it will be enhanced". This means that their metabolic processes must ultimately decrease the volume of the organism to some degree.

Most fish that have evolved in this harsh environment are not capable of surviving in laboratory conditions, and attempts to keep them in captivity have led to their deaths. Deep-sea organisms contain gas-filled spaces (vacuoles). Gas is compressed under high pressure and expands under low pressure. Because of this, these organisms have been known to blow up if they come to the surface. Other complications arise from nitrogen narcosis and decompression sickness, which also occur in humans. Nitrogen narcosis occurs because the absorption of gases in the blood, especially nitrogen, increase at greater depths. The result is similar to drunkenness. Decompression sickness occurs when excess gases cannot be removed from the blood stream fast enough when an organism rises in the water column. The decreased pressure makes the gases expand and small bubbles of nitrogen form in the blood stream as well as tissues. The result of this can be bone damage, extreme pain, physical debilitation, and even death.

This can be seen in the case of a frilled shark found in shallow waters near Japan. Frilled sharks usually live at a depth of 1,500 metres, and when this specimen was transferred to a marine park it died within a few hours. For this reason little is known about them, as there are limits to the amount of useful research that can be carried out on dead specimens and deep-sea exploratory equipment is very expensive. As such, many species are known only to scientists and only by scientific names.

## Characteristics

The fish of the deep-sea are among the strangest and most elusive creatures on Earth. In this deep unknown lie many unusual creatures that have yet to be studied. Since many of these fish live in regions where there is no natural illumination, they cannot rely solely on their eyesight for locating prey and mates and avoiding predators; deep-sea fish have evolved appropriately to the extreme sub-photic region in which they live. Many of these organisms are blind and rely on their other senses, such as sensitivities to changes in local pressure and smell, to catch their food and avoid being caught. Those that aren't blind have large and sensitive eyes that can use bioluminescent light. These eyes can be as much as 100 times more sensitive to light than human eyes. Also, to avoid predation, many species are dark to blend in with their environment.

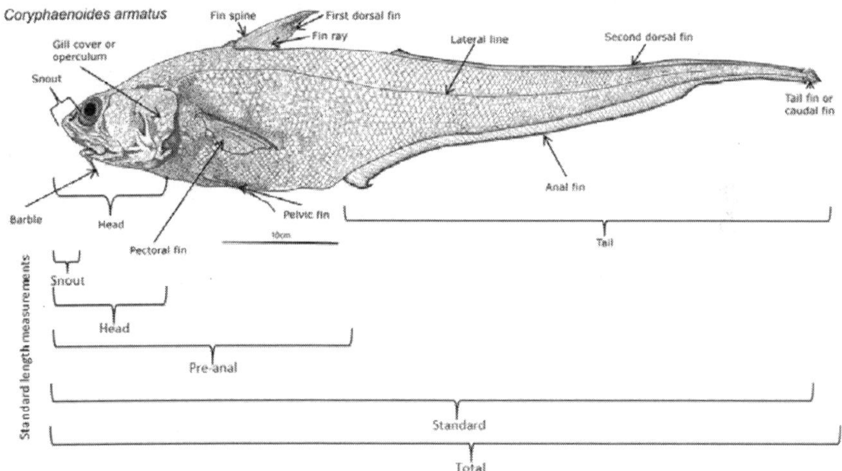


***Figure:*** *An annotated diagram of the basic external features of an Abyssal grenadier and standard length measurements.*

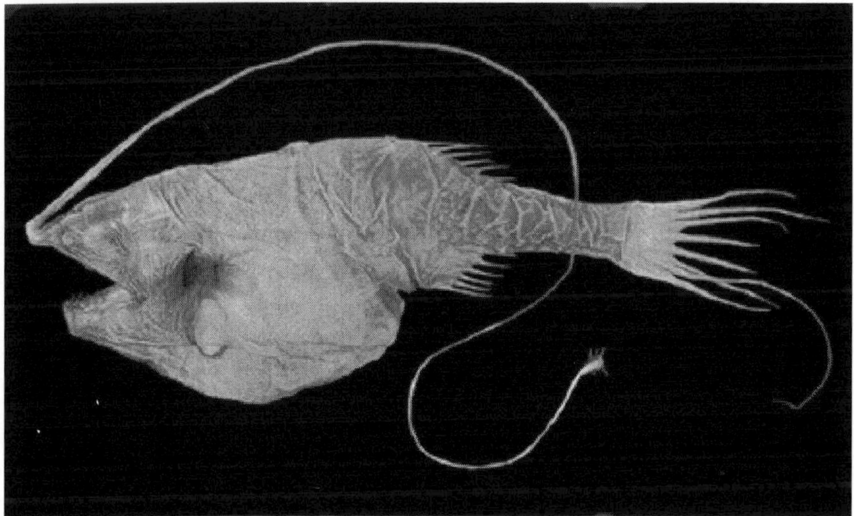

***Figure:*** *Gigantactis is a deep-sea fish with a dorsal fin whose first filament has become very long and is tipped with a bioluminescent photophore lure.*

***Figure:*** *Bigeye tuna cruise the epipelagic zone at night and the mesopelagic zone during the day.*

Many deep-sea fish are bioluminescent, with extremely large eyes adapted to the dark. Bioluminescent organisms are capable of producing light biologically through the agitation of molecules of luciferin, which then produce photons of light. This process must be done in the presence of oxygen. These organisms are common in the mesopelagic region and below (200m and below). More than 50% of deep-sea fish as well as some species of shrimp and squid are capable of bioluminescence. About 80% of these organisms have photophores – light producing glandular cells that contain luminous bacteria bordered by dark colourings. Some of these photophores contain lenses, much like those in the eyes of humans, which can intensify or lessen the emanation of light. The ability to produce light only requires 1% of the organism's energy and has many purposes: It is used to search for food and attract prey, like the anglerfish; claim territory through patrol; communicate and find a mate; and distract or temporarily blind predators to escape. Also, in the mesopelagic where some light still penetrates, some organisms camouflage themselves from predators below them by illuminating their bellies to match the colour and intensity of light from above so that no shadow is cast. This tactic is known as counter illumination.

The lifecycle of deep-sea fish can be exclusively deep water although some species are born in shallower water and sink upon maturation. Regardless of the depth where eggs and larvae reside, they are typically pelagic. This planktonic – drifting – lifestyle requires neutral buoyancy. In order to maintain this, the eggs and larvae often contain oil droplets in their plasma. When these organisms are in their fully matured state they need other adaptations to maintain their positions in the water column. In general, water's density causes upthrust – the aspect of buoyancy that makes organisms float. To counteract this, the density of an organism must be greater than that of the surrounding water. Most animal tissues are denser than water, so they must find an equilibrium to make them float. Many organisms develop swim bladders (gas cavities) to stay afloat, but because of the high pressure of their environment, deep-sea fishes usually do not have this organ. Instead they exhibit structures similar to hydrofoils in order to provide hydrodynamic lift. It has also been found that the deeper a fish lives, the more jelly-like its flesh and the more minimal its bone structure. They reduce their tissue density through high fat content, reduction of skeletal weight – accomplished through reductions of size, thickness, and mineral content – and water accumulation makes them slower and less agile than surface fish.

Due to the poor level of photosynthetic light reaching deep-sea environments, most fish need to rely on organic matter sinking from higher levels, or, in rare cases, hydrothermal vents for nutrients. This makes the deep-sea much poorer in productivity than shallower regions. Also, animals in the pelagic environment are sparse and food doesn't come along frequently. Because of this, organisms need adaptations that allow them to survive. Some have long feelers to help them locate prey or attract mates in the pitch black of the deep ocean. The deep-sea angler fish in particular has a long fishing-rod-like adaptation protruding from its face, on the end of which is a bioluminescent piece of skin that wriggles like a worm to lure its prey. Some must consume other fish that are the same size or larger than them and they need adaptations to help digest them efficiently. Great sharp teeth, hinged jaws, disproportionately large mouths, and expandable bodies are a few of the characteristics that deep-sea fishes have for this purpose. The gulper eel is one example of an organism that displays these characteristics.

Fish in the different pelagic and deep water benthic zones are physically structured, and behave in ways, that differ markedly from each other. Groups of coexisting species within each zone all seem to operate in similar ways, such as the small mesopelagic vertically migrating plankton-feeders, the bathypelagic anglerfishes, and the deep water benthic rattails. "

Ray finned species, with spiny fins, are rare among deep sea fishes, which suggests that deep sea fish are ancient and so well adapted to their environment that invasions by more modern fishes have been unsuccessful. The few ray fins that do exist are mainly in the Beryciformes and Lampriformes, which are also ancient forms. Most deep sea pelagic fishes belong to their own orders, suggesting a long evolution in deep sea environments. In contrast, deep water benthic species, are in orders that include many related shallow water fishes.

## Mesopelagic Fish

***Figure:*** *Most mesopelagic fishes are small filter feeders which ascend at night to feed in the nutrient rich waters of the epipelagic zone. During the day, they return to the dark, cold, oxygen deficient waters of the mesopelagic where they are relatively safe from predators. Lanternfish account for as much as 65 percent of all deep sea fish biomass and are largely responsible for the deep scattering layer of the world's oceans.*

***Figure:*** *Most of the rest of the mesopelagic fishes are ambush predators, like this sabertooth fish. The sabertooth which uses its telescopic, upward-pointing eyes to pick out prey silhouetted against the gloom above. Their recurved teeth prevent a captured fish from backing out.*

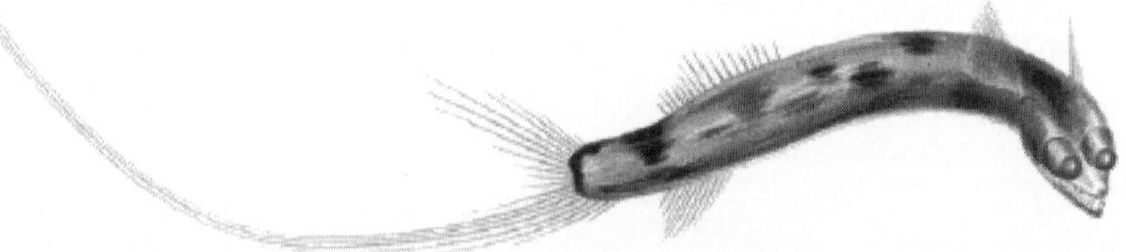

***Figure:*** *The telescopefish has large, forward-pointing telescoping eyes with large lenses.*

***Figure:*** *The Antarctic toothfish have large, upward looking eyes, adapted to detecting the silhouettes of prey fish.*

***Figure:*** *The Barreleye has barrel-shaped, tubular eyes which are generally directed upwards but can be swivelled forward.*

Below the epipelagic zone, conditions change rapidly. Between 200 metres and about 1000 metres, light continues to fade until there is almost none. Temperatures fall through a thermocline to temperatures between 39°F (3.9°C) and 46°F (7.8°C). This is the twilight or mesopelagic zone. Pressure continues to increase, at the rate of one atmosphere every 10 metres, while nutrient concentrations fall, along with dissolved oxygen and the rate at which the water circulates."

Sonar operators, using the newly developed sonar technology during World War II, were puzzled by what appeared to be a false sea floor 300–500 metres deep at day, and less deep at night. This turned out to be due to millions of marine organisms, most particularly small mesopelagic fish, with swimbladders that reflected the sonar. These organisms migrate up into shallower water at dusk to feed on plankton. The layer is deeper when the moon is out, and can become shallower when clouds pass over the moon. This phenomenon has come to be known as the deep scattering layer. Most mesopelagic fish make daily vertical migrations, moving at night into the epipelagic zone, often following similar migrations of zooplankton, and returning to the depths for safety during the day. These vertical migrations often occur over a large vertical distances, and are undertaken with the assistance of a swimbladder.

The swimbladder is inflated when the fish wants to move up, and, given the high pressures in the messoplegic zone, this requires significant energy. As the fish ascends, the pressure in the swimbladder must adjust to prevent it from bursting. When the fish wants to return to the depths, the swimbladder is deflated. Some mesopelagic fishes make daily migrations through the thermocline, where the temperature changes between 50°F (10°C) and 69°F (20°C), thus displaying considerable tolerances for temperature change.

These fish have muscular bodies, ossified bones, scales, well developed gills and central nervous systems, and large hearts and kidneys. Mesopelagic plankton feeders have small mouths with fine gill rakers, while the piscivores have larger mouths and coarser gill rakers. The vertically migratory fish have swimbladders. Mesopelagic fish are adapted for an active life under low light conditions. Most of them are visual predators with large eyes. Some of the deeper water fish have tubular eyes with big lenses and only rod cells that look upwards. These give binocular vision and great sensitivity to small light signals. This adaptation gives improved terminal vision at the expense of lateral vision, and allows the predator to pick out squid, cuttlefish, and smaller fish that are silhouetted against the gloom above them.

Mesopelagic fish usually lack defencive spines, and use colour to camouflage themselves from other fish. Ambush predators are dark, black or red. Since the longer, red, wavelengths of light do not reach the deep sea, red effectively functions the same as black. Migratory forms use countershaded silvery colours. On their bellies, they often display photophores producing low grade light. For a predator from below, looking upwards, this bioluminescence camouflages the silhouette of the fish. However, some of these predators have yellow lenses that filter the (red deficient) ambient light, leaving the bioluminescence visible.

The brownsnout spookfish, a species of barreleye, is the only vertebrate known to employ a mirror, as opposed to a lens, to focus an image in its eyes. Sampling via deep trawling indicates that lanternfish account for as much as 65% of all deep sea fish biomass.

Indeed, lanternfish are among the most widely distributed, populous, and diverse of all vertebrates, playing an important ecological role as prey for larger organisms. The estimated global biomass of lanternfish is 550 - 660 million metric tonnes, several times the entire world fisheries catch. Lanternfish also account for much of the biomass responsible for the deep scattering layer of the world's oceans. Sonar reflects off the millions of lanternfish swim bladders, giving the appearance of a false bottom.

Bigeye tuna are an epipelagic/mesopelagic species that eats other fish. Satellite tagging has shown that bigeye tuna often spend prolonged periods cruising deep below the surface during the daytime, sometimes making dives as deep as 500 metres. These movements are thought to be in response to the vertical migrations of prey organisms in the deep scattering layer.

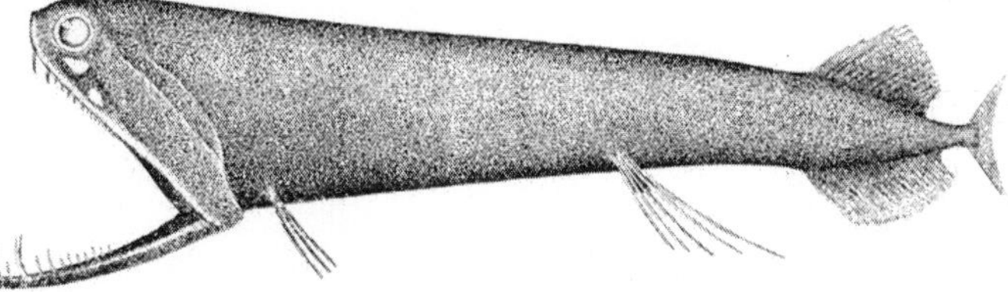

***Figure:*** *The stoplight loosejaw has a lower jaw one-quarter as long as its body. The jaw has no floor and is attached only by a hinge and a modified tongue bone. Large fang-like teeth in the front are followed by many small barbed teeth.*

***Figure:*** *The stoplight loosejaw is also one of the few fishes that produce red bioluminescence. As most of their prey cannot perceive red light, this allows it to hunt with an essentially invisible beam of light.*

***Figure:*** *Longnose lancetfish. Lancetfish are ambush predators which spend all their time in the mesopelagic zone. They are among the largest mesopelagic fishes (up to 2 metres).*

***Figure:*** *The daggertooth paralyses other mesopelagic fish when it bites them with its dagger-like teeth.*

## Bathypelagic Fish

### *Bathypelagic Fish*

***Figure:*** *The Sloane's viperfish can make nightly migrations from bathypelagic depths to near surface waters.*

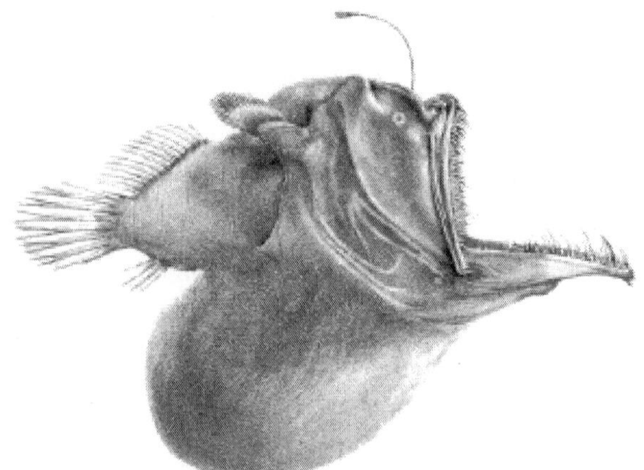

***Figure:*** *The humpback anglerfish is a bathypelagic ambush predator, which attracts prey with a bioluminescent lure. It can ingest prey larger than itself, which it swallows with an inrush of water when it opens its mouth.*

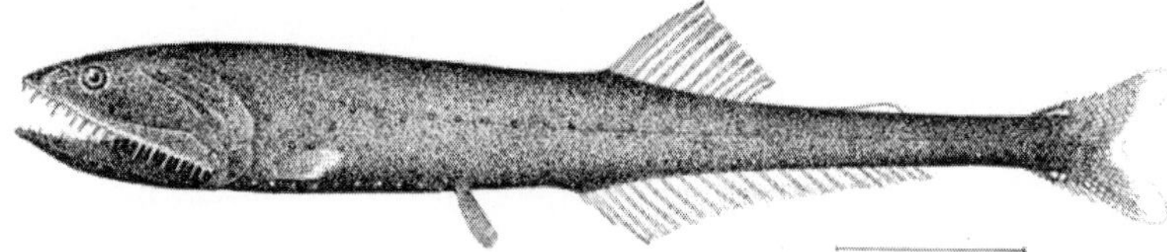

***Figure:*** *Many bristlemouth species, such as the "spark anglemouth" above, are also bathypelagic ambush predators which can swallow prey larger than themselves. They are among the most abundant of all vertebrate families.*

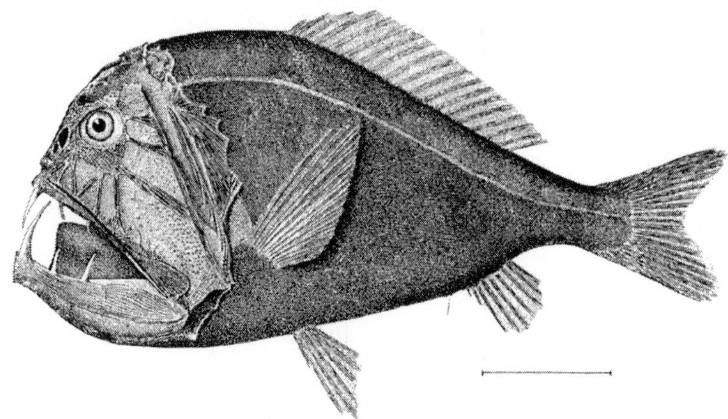

***Figure:*** *The widespread fangtooth has the largest teeth of any fish, proportionate to body size. Despite their ferocious appearance, bathypelagic fish are usually weakly muscled and too small to represent any threat to humans.*

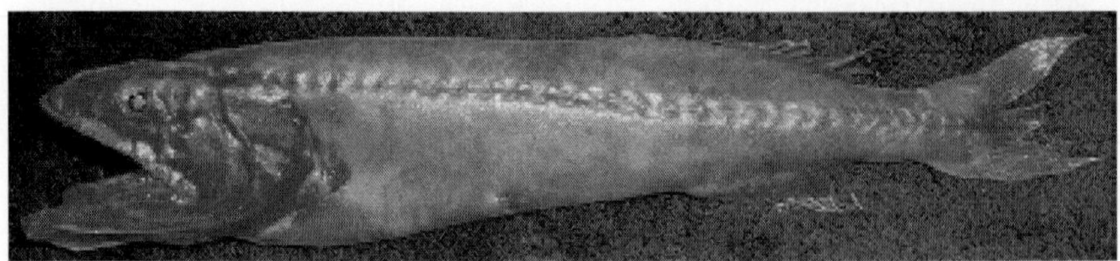

***Figure:*** *Young, red flabby whalefish make nightly vertical migrations into the lower mesopelagic zone to feed on copepods. When males make the transition to adults, they develop a massive liver, and then their jaws fuse shut. They no longer eat, but continue to metabolise the energy stored in their liver.*

Below the mesopelagic zone it is pitch dark. This is the midnight or bathypelagic zone, extending from 1000 metres to the bottom deep water benthic zone. If the water is exceptionally deep, the pelagic zone below 4000 metres is sometimes called the lower midnight or abyssopelagic zone. Conditions are somewhat uniform throughout these zones, the darkness is complete, the pressure is crushing, and temperatures, nutrients and dissolved oxygen levels are all low.

Bathypelagic fish have special adaptations to cope with these conditions – they have slow metabolisms and unspecialised diets, being willing to eat anything that comes along. They prefer to sit and wait for food rather than waste energy searching for it. The behaviour of bathypelagic fish can be contrasted with the behaviour of mesopelagic fish. Mesopelagic fish are often highly mobile, whereas bathypelagic fish are almost all lie-in-wait predators, normally expending little energy in movement.

The dominant bathypelagic fishes are small bristlemouth and anglerfish; fangtooth, viperfish, daggertooth and barracudina are also common. These fishes are small, many about 10 centimetres long, and not many longer than 25 cm. They spend most of their time waiting patiently in the water column for prey to appear or to be lured by their phosphors. What little energy is available in the bathypelagic zone filters from above in the form of detritus, faecal material, and the occasional invertebrate or mesopelagic fish. About 20 percent of the food that has its origins in the epipelagic zone falls down to the mesopelagic zone, but only about 5 percent filters down to the bathypelagic zone.

Bathypelagic fish are sedentary, adapted to outputting minimum energy in a habitat with very little food or available energy, not even sunlight, only bioluminescence. Their bodies are elongated with weak, watery muscles and skeletal structures. Since so much of the fish is water, they are not compressed by the great pressures at these depths.

They often have extensible, hinged jaws with recurved teeth. They are slimy, without scales. The central nervous system is confined to the lateral line and olfactory systems, the eyes are small and may not function, and gills, kidneys and hearts, and swimbladders are small or missing.

These are the same features found in fish larvae, which suggests that during their evolution, bathypelagic fish have acquired these features through neoteny. As with larvae, these features allow the fish to remain suspended in the water with little expenditure of energy. Despite their ferocious appearance, these beasts of the deep are mostly miniature fish with weak muscles, and are too small to represent any threat to humans.

The swimbladders of deep sea fish are either absent or scarcely operational, and bathypelagic fish do not normally undertake vertical migrations. Filling bladders at such great pressures incurs huge energy costs. Some deep sea fishes have swimbladders which function while they are young and inhabit the upper epipelagic zone, but they wither or fill with fat when the fish move down to their adult habitat.

The most important sensory systems are usually the inner ear, which responds to sound, and the lateral line, which responds to changes in water pressure. The olfactory system can also be important for males who find females by smell. Bathypelagic fish are black, or sometimes red, with few photophores. When photophores are used, it is usually to entice prey or attract a mate. Because food is so scarce, bathypelagic predators are not selective in their feeding habits, but grab whatever come close enough. They accomplish this by having a large mouth with sharp teeth for grabbing large prey and overlapping gill rakers which prevent small prey that have been swallowed from escaping.

It is not easy finding a mate in this zone. Some species depend on bioluminescence. Others are hermaphrodites, which doubles their chances of producing both eggs and sperm when an encounter occurs. The female anglerfish releases pheromones to attract tiny males. When a male finds her, he bites on to her and never lets go. When a male of the anglerfish species *Haplophryne mollis* bites into the skin of a female, he releases an enzyme that digests the skin of his mouth and her body, fusing the pair to the point where the two circulatory systems join up. The male then atrophies into nothing more than a pair of gonads. This extreme sexual dimorphism ensures that, when the female is ready to spawn, she has a mate immediately available.

Many forms other than fish live in the bathypelagic zone, such as squid, large whales, octopuses, sponges, brachiopods, sea stars, and echinoids, but this zone is difficult for fish to live in.

***Figure:*** *Flashlight fish have a retroreflector behind the retina which they use with photophores to detect eyeshine in other fish.*

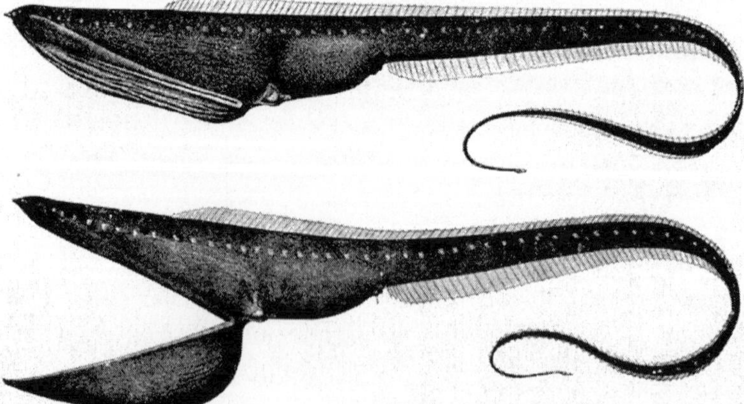

***Figure:*** *The gulper eel uses its mouth like a net by opening its large mouth and swimming at its prey. It has a luminescent organ at the tip of its tail to attract prey.*

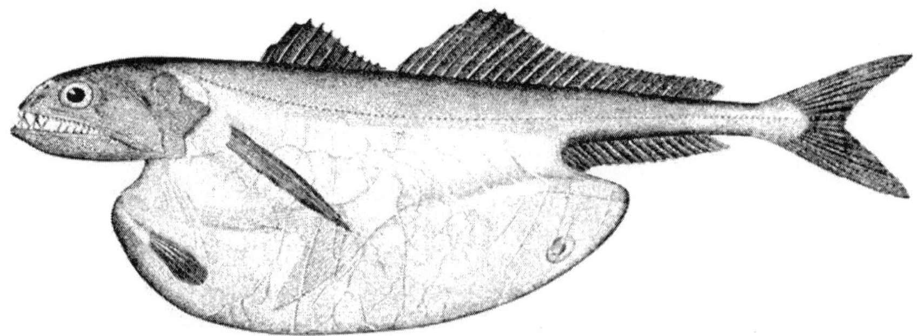

***Figure:** The black swallower, with its distensible stomach, is notable for its ability to swallow, whole, bony fishes ten times its mass.*

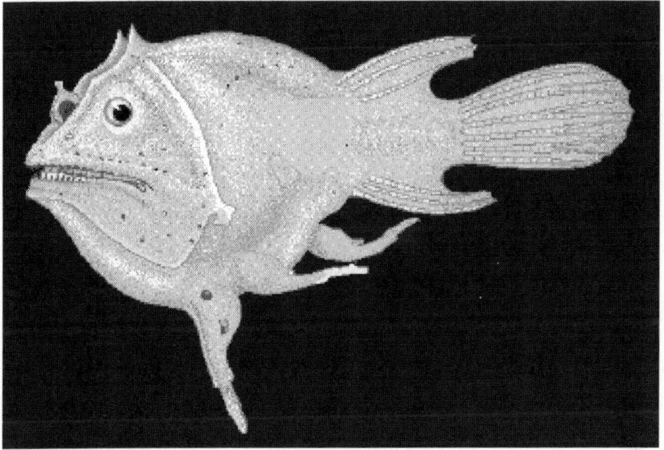

***Figure:** Female* Haplophryne mollis *anglerfish trailing attached males which have atrophied into a pair of gonads, for use when the female is ready to spawn.*

### Lanternfish

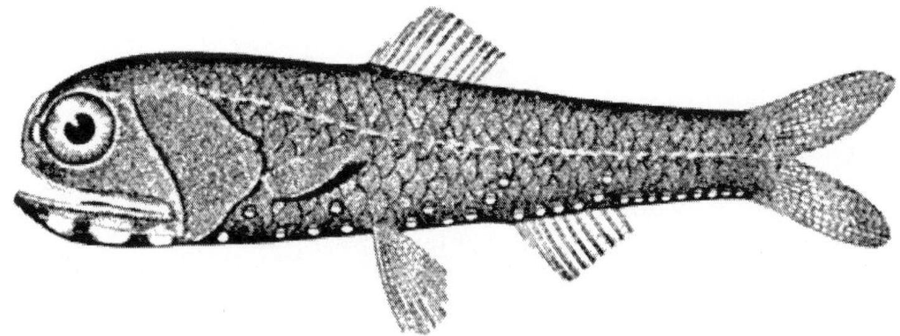

***Figure:** Lanternfish*

Sampling via deep trawling indicates that lanternfish account for as much as 65% of all deep-sea fish biomass. Indeed, lanternfish are among the most widely distributed, populous, and diverse of all vertebrates, playing an important ecological role as prey for larger organisms. With an estimated global biomass of 550 - 660 million metric tons, several times the entire world fisheries catch, lanternfish also account for much of the biomass responsible for the deep scattering layer of the world's oceans.

In the Southern Ocean, Myctophids provide an alternative food resource to krill for predators such as squid and the King Penguin. Although these fish are plentiful and prolific, currently only a few commercial lanternfish fisheries exist: These include limited operations off South Africa, in the sub-Antarctic, and in the Gulf of Oman.

### Endangered Species

A 2006 study by Canadian scientists has found five species of deep-sea fish – blue hake, spiny eel – to be on the verge of extinction due to the shift of commercial fishing from continental shelves to the slopes of the continental shelves, down to depths of 1600 metres. The slow reproduction of these fish – they reach sexual maturity at about the same age as human beings – is one of the main reasons that they cannot recover from the excessive fishing.

## Carp Fish

Carp are various species of oily freshwater fish of the family Cyprinidae, a very large group of fish native to Europe and Asia.

The cypriniformes (family Cyprinidae) are traditionally grouped with the Characiformes, Siluriformes and Gymnotiformes to create the superorder Ostariophysi, since these groups have certain common features, such as being found predominantly in fresh water and that they possess Weberian ossicles (an anatomical structure originally made up of small pieces of bone formed from four or five of the first vertebrae); the most anterior bony pair is in contact with the extension of the labyrinth and the posterior with the swimbladder.

The function is poorly understood, but this structure is presumed to take part in the transmission of vibrations from the swimbladder to the labyrinth and in the perception of sound, which would explain why the Ostariophysi have such a great capacity for hearing.

Most cypriniformes have scales and teeth on the inferior pharyngeal bones which may be modified in relation to the diet. *Tribolodon* is the only cyprinid genus which tolerates salt water, although there are several species which move into brackish water, but return to fresh water to spawn. All of the other cypriniformes live in continental waters and have a wide geographical range. Some consider all cyprinid fishes carp, and the family Cyprinidae itself is often known as the carp family.

In colloquial use, carp usually refers only to several larger cyprinid species such as *Cyprinus carpio* (common carp), *Carassius carassius* (Crucian carp), *Ctenopharyngodon idella* (grass carp), *Hypophthalmichthys molitrix* (silver carp), and *Hypophthalmichthys nobilis* (bighead carp). Carp have long been an important food fish to humans, as well as popular ornamental fishes such as the various goldfish breeds and the domesticated common carp variety known as koi.

As a result, carp have been introduced to various locations, though with mixed results. Several species of carp are listed as invasive species by the U.S. Department of Agriculture, and worldwide large sums of money are spent on carp control.

## Species

In 1653 Izaak Walton wrote in *The Compleat Angler*, "The Carp is the queen of rivers; a stately, a good, and a very subtle fish; that was not at first bred, nor hath been long in England, but is now naturalised."

*Some prominent carp in the family Cyprinidae*

| *Common name* | *Scientific name* | *Maximum length* | *Common length* | *Maximum weight* | *Maximum age* | *Trophic level* |
|---|---|---|---|---|---|---|
| Silver carp | *Hypophthalmichthys molitrix* (Valenciennes, 1844) | 105 cm | 18 cm | 50 kg | years | 2.0 |
| Common carp | *Cyprinus carpio* Linnaeus, 1758 | 110 cm | 31 cm | 40.1 kg | 38 years | 3.0 |
| Grass carp | *Ctenopharyngodon idella* (Valenciennes, 1844) | 150 cm | 10.7 cm | 45.0 kg | 21 years | 2.0 |
| Bighead carp | *Hypophthalmichthys nobilis* (Richardson, 1845) | 146 cm | 60 cm | 40.0 kg | 20 years | 2.3 |
| Crucian carp | *Carassius carassius* (Linnaeus, 1758) | 64 cm | 15 cm | 3.0 kg | 10 years | 3.1 |
| Catla carp (Indian carp) | *Cyprinus catla* (Hamilton, 1822) | 182 cm | cm | 38.6 kg | years | 2.8 |
| Mrigal carp | *Cirrhinus cirrhosus* (Bloch, 1795) | 100 cm | 40 cm | 12.7 kg | years | 2.5 |
| Black carp | *Mylopharyngodon piceus* (Richardson, 1846) | 122 cm | 12.2 cm | 35 kg | 13 years | 3.2 |
| Mud Carp | *Cirrhinus molitorella* (Valenciennes, 1844) | 55.0 cm | 15.2 cm | 0.50 kg | years | 2.0 |

*Recreational Fishing*

Carp are variable in terms of angling value.

- In Europe, even when not fished for food, they are eagerly sought by anglers, being considered highly prized coarse fish that are difficult to hook. The UK has a thriving carp angling market. It is the fastest growing angling market in the UK, and has spawned a number of specialised carp angling publications such as *Carpology*, *Advanced carp fishing*, *Carpworld* and *Total Carp*, and informative carp angling web sites, such as Carpfishing UK.

***Figure:*** *An angler with 17 kg mirror carp (*Cyprinus carpio*)*

- In the United States, carp are also classified as a rough fish, as well as damaging to naturalized exotic species, but with sporting qualities. Many states' departments of natural resources are beginning to view the carp as an angling fish instead of a maligned pest. Groups such as CarpPro, Wild Carp Companies, American Carp Society and the Carp Anglers Group promote the sport and work with fisheries departments to organise events to introduce and expose others to the unique opportunity the carp offers freshwater anglers.

### *Aquaculture*

Various species of carp have been domesticated and reared as food fish across Europe and Asia for thousands of years. These various species appear to have been domesticated independently, as the various domesticated carp species are native to different parts of Eurasia. Aquaculture has been pursued in China for at least 2,400 years. A tract by Fan Li in the fifth century BC details many of the ways carp were raised in ponds. The common carp, *Cyprinus carpio*, is originally from Central Europe. Several carp species (collectively known as Asian carp) were domesticated in East Asia. Carp that are originally from South Asia, for example catla (*Gibelion catla*), rohu (*Labeo rohita*) and mrigal (*Cirrhinus cirrhosus*), are known as Indian carp.

Their hardiness and adaptability have allowed domesticated species to be propagated all around the world.

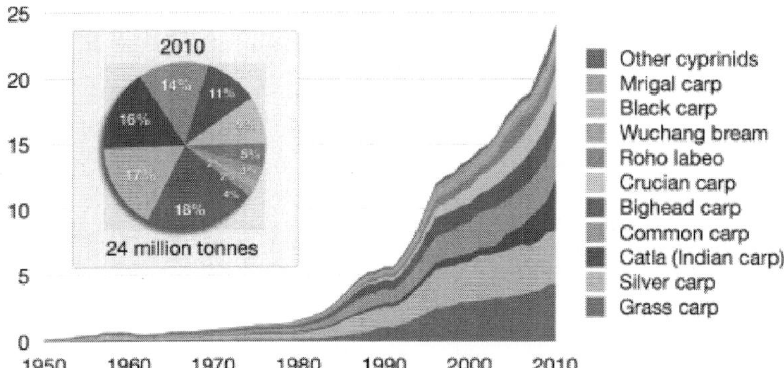


***Figure:*** *Aquaculture production of cyprinids by species in million tonnes, 1950–2010, as reported by the FAO.*

Although the carp was an important aquatic food item, as more fish species have become readily available for the table, the importance of carp culture in Western Europe has become less important. Demand has declined, partly due to the appearance of more desirable table fish such as trout and salmon through intensive farming, and environmental constraints. However, fish production in ponds is still a major form of aquaculture in Central and Eastern Europe, including the Russian Federation, where most of the production comes from low or intermediate-intensity ponds. In Asia, the farming of carp continues to surpass the total amount of farmed fish volume of intensively sea-farmed species, such as salmon and tuna.

## Breeding

Selective breeding programs for the Common carp (*Cyprinus carpio*) include improvement in growth, shape and resistance to disease. Experiments carried out in the USSR used crossings of broodstocks to increase genetic diversity and then selected the species for traits like growth rate, exterior traits and viability, and/or adaptation to environmental conditions like variations in temperature. selected carp for fast growth and tolerance to cold, the Ropsha carp. The results showed a 30-40% to 77.4% improvement of cold tolerance but did not provide any data for growth rate. An increase in growth rate was observed in the second generation in Vietnam, Moav and Wohlfarth (1976) showed positive results when selecting for slower growth for three generations compared to selecting for faster growth. Schaperclaus (1962) showed resistance to the dropsy disease wherein selected lines suffered low mortality (11.5%) compared to unselected (57%).

The major carp species used traditionally in Chinese aquaculture are the black, grass, silver and bighead carp. In the 1950s, the Pearl River Fishery Research Institute in China made a technological breakthrough in the induced breeding of these carps, which has resulted in a rapid expansion of freshwater aquaculture in China. In the late 1990s, scientists at the Chinese Academy of Fishery Sciences developed a new variant of the common carp called the Jian carp. This succulent fish grows rapidly and has a high feed conversion rate. Over 50% of the total aquaculture production of carp in China has now converted to Jian carp.

## As Food

Packaged grass carp fillets for sale

- Bighead carp is enjoyed in many parts of the world, but it has not become a popular foodfish in North America. Acceptance there has been hindered in part by the name "carp", and its association with the common carp which is not a generally favoured foodfish in North America. The flesh of the bighead carp is white and firm, different to that of the common carp, which is darker and richer. Bighead carp flesh does share one unfortunate similarity with common carp flesh - both have intramuscular bones within the filet. However, bighead carp captured from the wild in the United States tend to be much larger than common carp, so the intramuscular bones are also larger and less problematic.
- Crucian carp is considered the best-tasting pan fish in Poland. It is known as (Polish: *karaœ*), and is served traditionally with sour cream (*karasie w œmietanie*). In Russia, this particular species is called "golden crucian", and is one of the fish used in a borscht recipe called *borshch s karasej* or *borshch s karasyami.*
- Mud carp – due to the low cost of production, the fish is mainly consumed by the poor and locally consumed; it is mostly sold alive for eating, but can be dried and salted. The fish is sometimes canned or processed as fish cakes, fish balls or dumplings. They can be found for retail sale within China.
- Chinese mud carp is an important food fish in Guangdong Province. It is also cultured in this area and Taiwan. Cantonese and Shunde cuisines often use this fish to make fish balls and dumplings. It can be used with *douchi* or Chinese fermented

black beans in a dish called fried dace with salted black beans. It can be served cooked with vegetables such as Chinese cabbage.

- Fisherman's soup
- Kuai
- Taramosalata
- Masgouf, a popular Iraqi dish consisting of seasoned, grilled carp

### *As Ornamental Fish*

Carp, along with many of their cyprinid relatives, are popular ornamental aquarium and pond fish. The two most notable ornamental carps are goldfish and koi. Goldfish and koi have advantages over most other ornamental fishes, in that they are tolerant of cold (they can survive in water temperatures as low as 4°C), can survive at low oxygen levels, and can tolerate low water quality.

Goldfish (*Carassius auratus*) were originally domesticated from the Prussian carp (*Carassius gibelio*), a dark greyish-brown carp native to Asia. They were first bred for colour in China over a thousand years ago. Due to selective breeding, goldfish have been developed into many distinct breeds, and are found in various colours, colour patterns, forms and sizes far different from those of the original carp. Goldfish were kept as ornamental fish in Japan for hundreds of years before being introduced to China in the 15th century, and to Europe in the late 17th century.

Koi are a domesticated subspecies of common carp (*Cyprinus carpio*) that have been selectively bred for colour. The common carp was introduced from China to Japan, where selective breeding of the common carp in the 1820s in the Niigata region resulted in koi. In Japanese culture, koi are treated with affection, and seen as good luck. They are popular in other parts of the world as outdoor pond fish.

## Carp Cultivation in Japan

Koi or more specifically *nishikigoi*, literally "brocaded carp"), are ornamental varieties of domesticated common carp (*Cyprinus carpio*) that are kept for decorative purposes in outdoor koi ponds or water gardens.

Koi varieties are distinguished by colouration, patterning, and scalation. Some of the major colours are white, black, red, yellow, blue, and cream. The most popular category of koi is the *Gosanke*, which is made up of the *Kohaku, Taisho Sanshoku,* and *Showa Sanshoku* varieties.

## Etymology

The word *koi* comes from Japanese, simply meaning "carp". It includes both the dull grey fish and the brightly coloured varieties. What are known as koi in English are referred to more specifically as *nishikigoi* in Japan (literally meaning "brocaded carp"). In Japanese, *koi* is a homophone for another word that means "affection" or "love"; koi are therefore symbols of love and friendship in Japan.

## History

Carp are a large group of fish originally found in Central Europe and Asia. Various carp species were originally domesticated in East Asia, where they were used as food fish. The ability of carp to survive and adapt to many climates and water conditions allowed the domesticated species to be propagated to many new locations, including Japan. Natural colour mutations of these carp would have occurred across all populations. Carp were first bred for colour mutations in China more than a thousand years ago, where selective breeding of the Prussian carp (*Carassius gibelio*) led to the development of the goldfish.

The common carp was aquacultured as a food fish at least as far back as the fifth century BC in China, and in the Roman Empire during the spread of Christianity in Europe. Common carp were bred for colour in Japan in the 1820s, initially in the town of Ojiya in the Niigata prefecture on the northeastern coast of Honshu island. By the 20th century, a number of colour patterns had been established, most notably the red-and-white *Kohaku*. The outside world was not aware of the development of colour variations in koi until 1914, when the Niigata koi were exhibited in the annual exposition in Tokyo. At that point, interest in koi exploded throughout Japan. The hobby of keeping koi eventually spread worldwide. They are now commonly sold in most pet stores, with higher-quality fish available from specialist dealers.

Extensive hybridization between different populations has muddled the historical zoogeography of the common carp. However, scientific consensus is that there are at least two subspecies of the common carp, one from Western Eurasia (*Cyprinus carpio carpio*) and another from East Asia (*Cyprinus carpio haematopterus*). One recent study on the mitochondrial DNA of various common carp indicate that koi are of the East Asian subspecies. However, another recent study on the mitochondrial DNA of koi have found that koi are descended from multiple lineages of common carp from both Western Eurasian and East Asian varieties. This could be the result of koi being bred from a mix of East Asian and Western Eurasian carp varieties, or being bred exclusively from East Asian varieties and being subsequently hybridized with Western Eurasian varieties (the butterfly koi is one known product of such a cross). Which is true has not been resolved.

It was from this handful of Koi breeds that all other Nishikigoi types were bred, with the exception of the Ogon variety (single coloured, metallic Koi) which wasn't developed until recently. The last development of this early time was a great breakthrough in Koi breeding and is still revered as one of the most traditional of Koi breeds. A tri-coloured Koi called a Taisho Sanshoku, more commonly known as the Sanke, was first seen during the Meiji era (1868-1912). Though it is not known who first developed this breed, the Sanke was exhibited for the first time in 1915, when the Koi was about 15 years old.

### *Varieties*

Koi varieties are distinguished by colouration, patterning, and scalation. Some of the major colours are white, black, red, yellow, blue, and cream. While the possible colours are virtually limitless, breeders have identified and named a number of specific categories. The most popular category is Gosanke, which is made up of the *Kohaku, Taisho Sanshoku,* and *Showa Sanshoku* varieties.

New koi varieties are still being actively developed. Ghost koi developed in the 1980s have become very popular in the United Kingdom; they are a hybrid of wild carp and Ogon koi, and are distinguished by their metallic scales. Butterfly koi (also known as longfin koi, or dragon carp), also developed in the 1980s, are notable for their long and flowing fins. They are hybrids of koi with Asian carp. Butterfly koi and ghost koi are considered by some to be not true *nishikigoi*.

The major named varieties include:

1. Kohaku
2. Taisho Sanke
3. Showa Sanke
4. Tanchô
5. Chagoi
6. Asagi

7. Utsurimono
8. Bekko
9. Goshiki
10. Shûsui
11. Kinginrin
12. Kawarimono
13. Ôgon
14. Kumonryû
15. Ochiba
16. Koromo
17. Hikari-moyomono
18. Kikokuryû
19. Kin-Kikokuryû
20. Ghost koi
21. Butterfly koi
22. Doitsu-goi

- *Kôhaku* (v) is a white-skinned koi, with large red markings on the top. The name means "red and white"; *kohaku* was the first ornamental variety to be established in Japan (late 19th century).
- *Taishô Sanshoku* (or *Taisho Sanke)* ('Yck Nr,) is very similar to the *kohaku*, except for the addition of small black markings called *sumi* (¨X). This variety was first exhibited in 1914 by the koi breeder Gonzo Hiroi, during the reign of the Taisho Emperor. In America, the name is often abbreviated to just "Sanke". The kanji, Nr,, may be read as either *sanshoku* or as *sanke*.
- *Shôwa Sanshoku* (or *Showa Sanke*) is a black koi with red and white markings. The first *Showa Sanke* was exhibited in 1927, during the reign of the Showa Emperor. In America, the name is often abbreviated to just "Showa". The amount of *shiroji* on *Showa Sanke* has increased in modern times, to the point that it can be difficult to distinguish from *Taisho Sanke*. The kanji, Nr,, may be read as either *sanshoku* or as *sanke*.
- *Tanchô* is any koi with a solitary red patch on its head. The fish may be a *Tancho Showa*, *Tancho Sanke*, or even

*Tancho Goshiki*. It is named for the Japanese crane (*Grus japonensis*), which also has a red spot on its head.

- *Chagoi*, "tea-coloured", this koi can range in colour from pale olive-drab green or brown to copper or bronze and more recently, darker, subdued orange shades. Famous for its docile, friendly personality and large size, it is considered a sign of good luck among koi keepers.
- *Asagi* koi is light blue above and usually red below, but also occasionally pale yellow or cream, generally below the lateral line and on the cheeks. The Japanese name means pale greenish-blue, spring onion colour, or indigo. Sometimes it is incorrectly written as EmÄž (light yellow).
- *Utsurimono* is a black koi with a white, red, or yellow markings, in a zebra colour pattern. The oldest attested form is the yellow form, called "black and white markings" in the 19th century, but renamed *Ki Utsuri* by Elizaburo Hoshino, an early 20th-century koi breeder. The red and white versions are called *Hi Utsuri* and *Shiro Utsuri*, respectively. The word *utsuri* means to print (the black markings are reminiscent of ink stains). Genetically, it is the same as *Showa*, but lacking either red pigment (*Shiro Utsuri*) or white pigment (*Hi Utsuri*/*Ki Utsuri*).
- *Bekko* is a white-, red-, or yellow-skinned koi with black markings *sumi*. The Japanese name means "tortoise shell," and is commonly written as y0c02u. The white, red, and yellow varieties are called *Shiro Bekko, Aka Bekko* and *Ki Bekko*, respectively. It may be confused with the *Utsuri*.
- *Goshiki* is a dark koi with red (*Kohaku* style) *hi* pattern. The Japanese name means "five colours". It appears similar to an *Asagi*, with little or no *hi* below the lateral line and a *Kohaku Hi* pattern over reticulated (fishnet pattern) scales. The base colour can range from nearly black to very pale, sky blue.
- *Shûsui* means "autumn green"; the *Shûsui* was created in 1910 by Yoshigoro Akiyamaÿ, by crossing Japanese *Asagi* with German mirror carp. The fish has no scales, except for a single line of large mirror scales dorsally, extending from head to tail. The most common type of *Shûsui* have a pale, sky-blue/gray colour above the lateral line and red or orange (and very, very rarely bright yellow) below the lateral line and on the cheeks.

- *Kinginrin* is a koi with metallic (glittering, metal-flake-appearing) scales. The name translates into English as "gold and silver scales"; it is often abbreviated to *Ginrin*. There are *Ginrin* versions of almost all other varieties of koi, and they are fashionable. Their sparkling, glittering scales contrast to the smooth, even, metallic skin and scales seen in the Ogon varieties. Recently, these characteristics have been combined to create the new *ginrin Ogon* varieties.
- *Kawarimono* is a "catch-all" term for koi that cannot be put into one of the other categories. This is a competition category, and many new varieties of koi compete in this one category. It is also known as *kawarigoi*.
- *Ôgon* is a metallic koi of one colour only. The most commonly encountered colours are gold, platinum, and orange. Cream specimens are very rare. *Ogon* compete in the *Kawarimono* category and the Japanese name means "gold." The variety was created by Sawata Aoki in 1946 from wild carp he caught in 1921. Recently, the metallic-skinned *Ogon* is being crossed with *ginrin*-scaled fish to create the *ginrin Ogon* with metallic skin and sparkling (metal flake) scales.
- *Kumonryû* ÿliterally "nine tattooed dragons" ÿ is a black *doitsu*-scaled fish with curling white markings. The patterns are thought to be reminiscent of Japanese ink paintings of dragons. They famously change colour with the seasons. *Kumonryu* compete in the *Kawarimono* category.
- *Ochiba* is a light blue/gray koi with copper, bronze, or yellow (*Kohaku*-style) pattern, reminiscent of autumn leaves on water. The Japanese name means "fallen leaves".
- *Koromo* is a white fish with a *Kohaku*-style pattern with blue or black-edged scales only over the *hi* pattern. This variety first arose in the 1950s as a cross between a *Kohaku* and an *Asagi*. The most commonly encountered *Koromo* is an *Ai Goromo*, which is coloured like a *Kohaku*, except each of the scales within the red patches has a blue or black edge to it. Less common is the *Budo-Goromo*, which has a darker (burgundy) *hi* overlay that gives it the appearance of bunches of grapes. Very rarely seen is the *Tsumi-Goromo* which is similar to *Budo-Goromo*, but the *hi* pattern is such a dark burgundy that it appears nearly black.

- *Hikari-moyomono* is a koi with coloured markings over a metallic base or in two metallic colours.
- *Kikokuryûÿ*", literally "sparkle" or "glitter black dragon" ÿ is a metallic-skinned version of the *Kumonryu.*
- *Kin-Kikokuryû,* literally "gold sparkle black dragon" or "gold glitter black dragon" ÿis a metallic-skinned version of the *Kumonryu* with a *Kohaku*-style *hi* pattern developed by Mr. Seiki Igarashi of Ojiya City. There are (at least) six different genetic subvarieties of this general variety.
- Ghost koiÿ, a hybrid of *Ogon* and wild carp with metallic scales, is considered by some to be not *nishikigoi.*
- Butterfly koiÿ-œw is a hybrid of koi and Asian carp with long flowing fins. Various colourations depend on the koi stock used to cross. It also is considered by some to not be *nishikigoi.*
- *Doitsu-goi* originated by crossbreeding numerous different established varieties with "scaleless" German carp (generally, fish with only a single line of scales along each side of the dorsal fin). Also written as ìr8□É›, there are four main types of *Doitsu* scale patterns. The most common type (referred to above) has a row of scales beginning at the front of the dorsal fin and ending at the end of the dorsal fin (along both sides of the fin). The second type has a row of scales beginning where the head meets the shoulder and running the entire length of the fish (along both sides). The third type is the same as the second, with the addition of a line of (often quite large) scales running along the lateral line (along the side) of the fish, also referred to as "mirror koi". The fourth (and rarest) type is referred to as "armor koi" and are completely (or nearly) covered with very large scales that resemble plates of armor. They also are called *Kagami-goi*ÿ, or mirror carpÿ.

### Differences from Goldfish

Goldfish were developed in China more than a thousand years ago by selectively breeding Prussian carp for colour mutations. By the Song Dynasty (960 – 1279), yellow, orange, white, and red-and-white colourations had been developed. Goldfish (*Carassius auratus*) and Prussian carp (*Carassius gibelio*) are now considered different species. Goldfish were introduced to Japan in the 16th century and to Europe in the 17th century. Koi, on the other hand, were developed from common carp in Japan in the 1820s. Koi are domesticated common carp (*Cyprinus carpio*) that are selected or culled for colour; they are not a different species, and will revert to the original colouration within a few generations if allowed to breed freely.

In general, goldfish tend to be smaller than koi, and have a greater variety of body shapes and fin and tail configurations. Koi varieties tend to have a common body shape, but have a greater variety of colouration and colour patterns. They also have prominent barbels on the lip. Some goldfish varieties, such as the common goldfish, comet goldfish, and shubunkin have body shapes and colouration that are similar to koi, and can be difficult to tell apart from koi when immature. Since goldfish and koi were developed from different species of carp, even though they can interbreed, their offspring are sterile.

### Health, Maintenance and Longevity

The common carp is a hardy fish, and koi retain that durability. Koi are cold-water fish, but benefit from being kept in the 15-25 °C (59-77°F) range, and do not react well to long, cold, winter temperatures; their immune systems "turn off" below 10°C. Koi ponds usually have a metre or more of depth in areas of the world that become warm during the summer, whereas in areas that have harsher winters, ponds generally have a minimum of 1.5 metres (4½ feet). Specific pond construction has evolved by koi keepers intent on raising show-quality koi.

Koi's bright colours put them at a severe disadvantage against predators; a white-skinned *Kohaku* is a visual dinner bell against the dark green of a pond. Herons, kingfishers, otters, raccoons, cats, foxes, badgers and hedgehogs are all capable of emptying a pond of its fish. A well-designed outdoor pond will have areas too deep for herons to stand, overhangs high enough above the water that mammals cannot reach in, and shade trees overhead to block the view of aerial passers-by. It may prove necessary to string nets or wires above the surface. A pond usually includes a pump and filtration system to keep the water clear.

Koi are an omnivorous fish, and will eat a wide variety of foods, including peas, lettuce, and watermelon. Koi food is designed not only to be nutritionally balanced, but also to float so as to encourage them to come to the surface. When they are eating, it is possible to check koi for parasites and ulcers. Koi will recognise the persons feeding them and gather around them at feeding times. They can be trained to take food from one's hand. In the winter, their digestive systems slow nearly to a halt, and they eat very little, perhaps no more than nibbles of algae from the bottom. Care should be taken by hobbyists that proper oxygenation and off-gassing occurs over the winter months in small water ponds, so they do not perish. Their appetites will not come back until the water becomes warm in the spring. When the temperature drops below 10°C (50°F), feeding, particularly with protein, is halted or the food can spoil in their stomachs, causing sickness.

One famous scarlet koi, named "Hanako", was owned by several individuals, the last of whom was Dr. Komei Koshihara. Hanako was supposedly 226 years old upon her death in 1977, based on examining one of her scales in 1966. Koi "maximum longevity" is listed as 47 years old.

### Disease

Koi are very hardy. With proper care, they resist many of the parasites that affect more sensitive tropical fish species, such as Trichodina, Epistylis, Ich and other ciliated protozoans. Two of the biggest health concerns among koi breeders are the koi herpes virus (KHV) and Rhabdovirus carpio, which causes spring viraemia of carp (SVC). No treatment exists for either disease. Only biosecurity measures such as prompt detection, isolation and disinfection of tanks and equipment can prevent the spread of the disease and limit the loss of fish stock. In 2002, spring viraemia struck an ornamental koi farm in Kernersville, North Carolina, and required complete depopulation of the ponds and a lengthy quarantine period. For a while after this, some koi farmers in neighbouring states stopped importing fish for fear of infecting their own stocks.

### Breeding

Like most fish, koi reproduce through spawning in which a female lays a vast number of eggs and one or more males fertilize them. Nurturing the resulting offspring (referred to as "fry") is a tricky and tedious job, usually done only by professionals. Although a koi breeder may carefully select the parents they wish based on their desired characteristics, the resulting fry will nonetheless exhibit a wide range of colour and quality.

Koi will produce thousands of offspring from a single spawning. However, unlike cattle, purebred dogs, or more relevantly, goldfish, the large majority of these offspring, even from the best champion-grade koi, will not be acceptable as *nishikigoi* (they have no interesting colours) or may even be genetically defective. These unacceptable offspring are culled at various stages of development based on the breeder's expert eye and closely guarded trade techniques. Culled fry are usually destroyed or used as feeder fish (mostly used for feeding arowana due to the belief it will enhance its colour), while older culls, within their first year between 3" to 6" long (also called "Tosai"), are often sold as lower-grade, pond-quality koi.

The semirandomized result of the koi's reproductive process has both advantages and disadvantages for the breeder. While it requires diligent oversight to narrow down the favourable result the breeder wants, it also makes possible the development of new varieties of koi within relatively few generations.

### *In the Wild*

Koi have been accidentally or deliberately released into the wild in every continent except Antarctica. In many areas, they are considered an invasive species and pests. They greatly increase the turbidity of the water because they are constantly stirring up the substrate. This makes waterways unattractive, reduces the abundance of aquatic plants, and can render the water unsuitable for swimming or drinking, even by livestock. In some countries, koi have caused so much damage to waterways that vast amounts of money and effort have been spent trying to eradicate them, largely unsuccessfully.

## Ilish in North-East India

Ilish, also spelled Elish, *Tenualosa ilisha*, is a popular fish to eat among the people of South Asia. A tropical fish, it is the most popular fish with Bengalis and Oriyas, the national fish of Bangladesh and extremely popular in parts of India such as West Bengal, Odisha, Tripura, Assam and Southern Gujarat. Ilish also can be found in India's Assamese-, Bengali-, Oriya- and Telugu-speaking regions and in Pakistan's Sindh province. In Gujarat it is known as either Modenn or Palva.

Each year a large number of fish are caught in the Padma-Meghna-Jamuna delta, which flows into the Bay of Bengal. It is a sea fish but it lay eggs in large rivers. After being born the young Ilish (known as Jatka) then swim back to the sea. They are caught before they swim to the sea. Ilish is also caught from the sea. However, those caught from the sea are not considered to be as tasty as those caught from the river. The fish is full of tiny bones which require trained eating/hands to handle.

In Southern Gujarat, Bharuch located on the banks of river Narmada is famous for this fish. The fish from Bharuch is in huge demand in Mumbai and is even exported to many foreign countries. The fish in coastal area of Gujarat is known as Modenn if it is female and Palva if it is young male. As it is anadromous in nature (an uncommon phenomenon in tropical waters), the Ilish lives in the sea for most of its life, but migrates up to 1,200 km inland through rivers in the Indian sub-continent for spawning. Distances of 50–100 km are usually normal in the Bangladesh rivers.

In Bangladesh, Ilish is mainly caught in the Padma (lower Ganges), Meghna (lower Brahmaputra), and Jamuna rivers. Those from the Padma are considered to be the best in taste. In India, the Rupnarayan (which has the *Kolaghater Ilish*), Ganges, Mahanadi, Chilka Lake, Narmada and Godavari rivers are famous for their tasty breeds. Ilish is also found in the deltaic region of southern Pakistan, in the province of Sindh. Here it is commonly referred to as the *Palla* fish. The fish was usually found in abundant quantities in the district of Thatta. Recently, however, the lower reaches of the Indus have dried up as water is stored upstream, and the *Palla* cannot make its journey into the river any more.

## As Food

***Figure:** Panta Ilish - a traditional platter of congee with fried Ilish slice, supplemented with dried fish (*Shutki*), pickles (*Achar*), dal, green chillies and onion - is a popular serving for the Pohela Boishakh festival.*

***Figure:*** *Ilishi maachha curry with ginger mustard garlic paste in tomato seasoning in Odisha style in Oriya cuisine.*

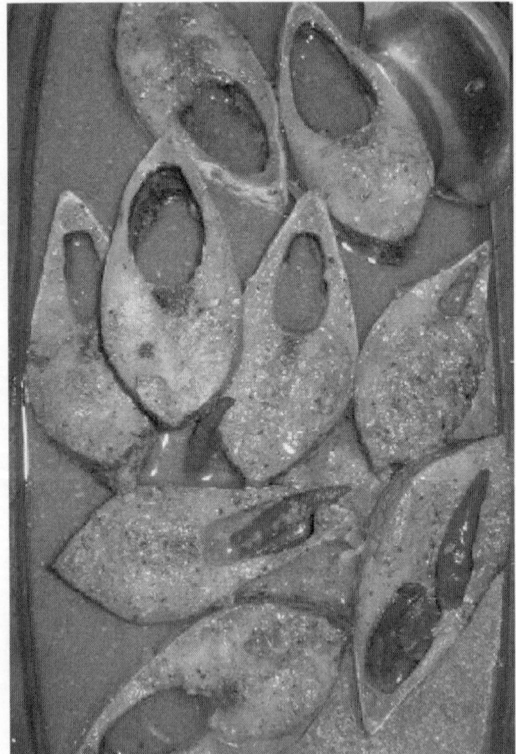

***Figure:*** Shorshe Ilish, *a dish of smoked ilish with mustard seeds, has been an important part of Bengali cuisine.*

Ilish is an oily fish rich in essential fatty acids (omega 3 fatty acids). Recent experiments have shown its beneficial effects in decreasing cholesterol level in rats and insulin level.

In Bengal, ilish can be smoked, fried, steamed, baked in young plantain leaves, prepared with mustard seed paste, curd, Begun (eggplant), different condiments like jira and so on. It is said that people can cook ilish in more than 50 ways. Ilish roe is also popular as a side dish. Ilish can be cooked in very little oil since the fish itself is very oily.

In North America (where Ilish is not always readily available) the shad fish is sometimes used as an Ilish substitute, especially in Bengali cuisine. This typically occurs near the East coast of North America, where fresh shad can be fished. The substitution is possible because of the fairly similar flavour and consistency of these two fish.

### Ilish in Culture

In many Bengali Hindu families a pair of Ilish fishes (Bengali: *Joda Ilish*) are bought on special auspicious days, like some pujas. It is considered auspicious to buy two Ilish fishes on the day of Saraswati Puja, which takes place in the beginning of Spring and also on the day of Lakshmi Puja (The Goddess of Wealth and Prosperity) which takes place in autumn. But this custom is prevalent mainly among the Bengali Hindus of former East Bengal many of whom now live in West Bengal, Barak Valley in Assam and Tripura in India after the Partition of India. Some of them give Ilish fish as an offering to the goddess Lakshmi, without which the Puja is sometimes thought to be incomplete.

In Odisha there is a popular saying that "Machha khaaiba Ilishii, chakiri kariba polisi", which means that eating Ilish and getting a job in Police department are of equal status.

### Ilish Production

Five type of ilish can be found worldwide. Yearly ilish caught are 5,000,000 ton. Among them, 50%-60% are caught by Bangladesh, 15%-20% are caught by India, Pakistan and rest 5%-10% are caught by Malaysia, Thailand, China, Vietnam and Srilanka.

### Overfishing and Possible Extinction

However, the ilish is rapidly becoming an overfished species, leading to a downward spiral of substantial price increases and collapsing populations. In the past Ilish were not harvested between Lakshmi Puja and Saraswati Puja due to informal customs of Bengali

Hindus. As disposable incomes have grown with economic advancement, many especially wealthier consumers have largely abandoned this practise. The resulting increase in prices has led to rampant overharvesting that has caused fish stocks to decline markedly in recent years, both in fish size and in catch size. With fewer and fewer fish to catch (and with the advent of finer fishing nets and more advanced trawling techniques), fishermen have been ignoring calls to at least leave the juvenile "jatka" alone to repopulate the species while traders are bidding up the price of the fish to exorbitant levels. Furthermore, the changes brought about by global warming have led to a gradual depletion of the ilish's breeding grounds, reducing populations further. The fish may be getting to the point of being functionally extinct from the Ganges Delta region as a result. While ilish are still available from as far away as India's Gujarat state and Pakistan, Burma, and parts of Southeast Asia, continued price increases is likely to put pressure on those populations as well so long as unscrupulous traders / fishermen and uncaring consumers do not change their behaviour.

## Catfish

Catfishes (order Siluriformes) are a diverse group of ray-finned fish. Named for their prominent barbels, which resemble a cat's whiskers, catfish range in size and behaviour from the heaviest and longest, the Mekong giant catfish from Southeast Asia and the second longest, the wels catfish of Eurasia, to detritivores (species that eat dead material on the bottom), and even to a tiny parasitic species commonly called the candiru, *Vandellia cirrhosa*. There are armour-plated types and also naked types, neither having scales. Despite their name, not all catfish have prominent barbels; members of the Siluriformes order are defined by features of the skull and swimbladder. Catfish are of considerable commercial importance; many of the larger species are farmed or fished for food. Many of the smaller species, particularly the genus *Corydoras*, are important in the aquarium hobby. Catfish are nocturnal.

### *Distribution and Habitat*

Extant catfish species live inland or in coastal waters of every continent except Antarctica. Catfish have inhabited all continents at one time or another. Catfish are most diverse in tropical South America, North America, Africa, and Asia. More than half of all catfish species live in the Americas. They are the only ostariophysans that have entered freshwater habitats in Madagascar, Australia, and New Guinea.

They are found in freshwater environments, though most inhabit shallow, running water. Representatives of at least eight families are hypogean (live underground) with three families that are also troglobitic (inhabiting caves). One such species is *Phreatobius cisternarum*, known to live underground in phreatic habitats. Numerous species from the families Ariidae and Plotosidae, and a few species from among the Aspredinidae and Bagridae, are found in salt water.

In the United States, catfish species may be known by a variety of slang names, such as "mud cat", "polliwogs", or "chuckleheads". These nicknames are not standardized, so one area may call a bullhead catfish by the nickname "chucklehead", while in another state or region, that nickname refers to the blue catfish.

## Physical Characteristics

### *External Anatomy of Catfish*

***Figure:*** *The armor plates are evident in* Corydoras semiaquilus.

Most catfish are bottom feeders. In general, they are negatively buoyant, which means that they will usually sink rather than float due to a reduced gas bladder and a heavy, bony head. Catfish have a variety of body shapes, though most have a cylindrical body with a flattened ventrum to allow for benthic feeding.

A flattened head allows for digging through the substrate as well as perhaps serving as a hydrofoil. Most have a mouth that can expand to a large size and contains no incisiform teeth; catfish generally feed through suction or gulping rather than biting and cutting prey. However, some families, notably Loricariidae and Astroblepidae, have a suckermouth that allows them to fasten themselves to objects in fast-moving water. Catfish also have a maxilla reduced to a support for barbels; this means that they are unable to protrude their mouths as other fish such as carp.

Catfish may have up to four pairs of barbels: nasal, maxillary (on each side of mouth), and two pairs of chin barbels, even though pairs of barbels may be absent depending on the species. Catfish also have chemoreceptors across their entire bodies, which means they "taste" anything they touch and "smell" any chemicals in the water. "In catfish, gustation plays a primary role in the orientation and location of food". Because their barbels and chemoreception are more important in detecting food, the eyes on catfish are generally small. Like other ostariophysans, they are characterized by the presence of a Weberian apparatus. Their well-developed Weberian apparatus and reduced gas bladder allow for improved hearing as well as sound production.

***Figure:*** *The channel catfish has four pairs of barbels.*

Catfish have no scales; their bodies are often naked. In some species, the mucus-covered skin is used in cutaneous respiration, where the fish breathes through its skin. In some catfish, the skin is covered in bony plates called scutes; some form of body armor appears in various ways within the order. In loricarioids and in the Asian genus *Sisor*, the armor is primarily made up of one or more rows of free dermal plates. Similar plates are found in large specimens of *Lithodoras*. These plates may be supported by vertebral processes, as in scoloplacids and in *Sisor*, but the processes never fuse to the plates or form any external armor. By contrast, in the subfamily Doumeinae (family Amphiliidae) and in hoplomyzontines (Aspredinidae), the armor is formed solely by expanded vertebral processes that form plates. Finally, the lateral armor of doradids, *Sisor*, and hoplomyzontines consists of hypertrophied lateral line ossicles with dorsal and ventral lamina.

All catfish, except members of Malapteruridae (electric catfish), possess a strong, hollow, bonified leading spine-like ray on their dorsal and pectoral fins. As a defence, these spines may be locked into place so that they stick outwards, which can inflict severe wounds. In several species catfish can use these fin rays to deliver a stinging protein if the fish is irritated. This venom is produced by glandular cells in the epidermal tissue covering the spines. In members of the family Plotosidae, and of the genus *Heteropneustes*, this protein is so strong it may hospitalize humans, those unfortunate enough to receive a sting; in *Plotosus lineatus*, the stings may result in death.

Juvenile catfish, like most fish, have relatively large heads, eyes and posterior median fins in comparison to larger, more mature individuals. These juveniles can be readily placed in their families, particularly those with highly derived fin or body shapes; in some cases identification of the genus is possible. As far as known for most catfish, features that are often characteristic of species such as mouth and fin positions, fin shapes, and barbel lengths show little difference between juveniles and adults. For many species, pigmentation pattern is also similar in juveniles and adults. Thus, juvenile catfishes generally resemble and develop smoothly into their adult form without distinct juvenile specialisations. Exceptions to this are the ariid catfishes, where the young retain yolk sacs late into juvenile stages, and many pimelodids, which may have elongated barbels and fin filaments or colouration patterns.

Sexual dimorphism is reported in about half of all families of catfish. The modification of the anal fin into an intromittent organ (in internal fertilizers) as well as accessory structures of the reproductive apparatus (in both internal and external fertilizers) have been described in species belonging to 11 different families.

### Size

Catfish have one of the greatest ranges in size within a single order of bony fish. Many catfish have a maximum length of under

12 cm. Some of the smallest species of Aspredinidae and Trichomycteridae reach sexual maturity at only 1 centimetre (0.39 in).

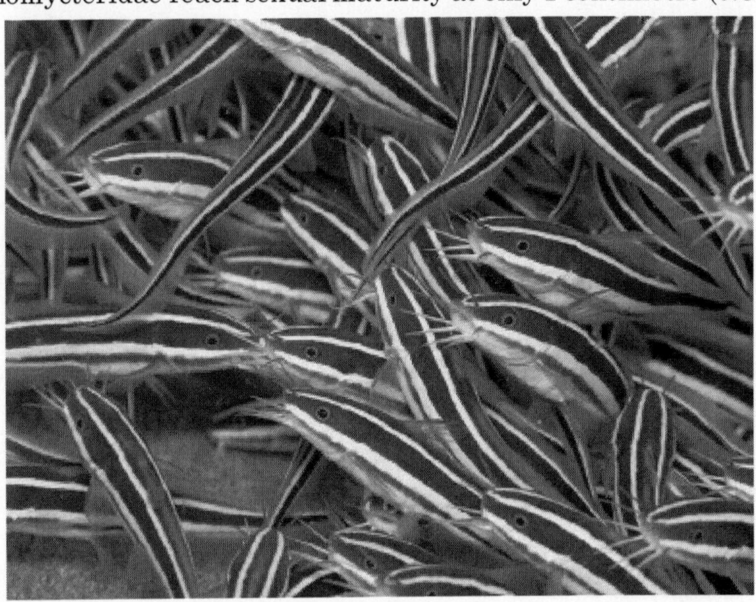

***Figure:*** *A sting from the striped eel catfish,* Plotosus lineatus, *may be fatal.*

The wels catfish, *Silurus glanis,* is the only native catfish species of Europe, besides the much smaller related Aristotle's catfish found in Greece. Mythology and literature record wels catfish of astounding proportions, yet to be proven scientifically. The average size of the species is about 1.2–1.6 m (3.9–5.2 ft), and fish more than 2 metres (6.6 ft) are very rare. The largest specimens on record measure more than 2.5 metres (8.2 ft) in length and sometimes exceeded 100 kilograms (220 lb).

The largest *Ictalurus furcatus,* caught in the Missouri River on July 20, 2010, weighed 130 pounds (59 kg). The largest flathead catfish, *Pylodictis olivaris,* ever caught was in Independence, Kansas, weighing 123 lb 9 oz (56.0 kg). In July 2009, a catfish weighing 193 pounds was caught in the River Ebro, Spain, by an 11-year old British schoolgirl. However, these records pale in comparison to a giant Mekong catfish caught in northern Thailand on May 1, 2005 and reported to the press almost 2 months later that weighed 293 kilograms (646 lb). This is the largest giant Mekong catfish caught since Thai officials started keeping records in 1981. The giant Mekong catfish are not well studied since they live in developing countries and it is quite possible that they can grow even larger.

### Internal Anatomy

In many catfish, the *humeral process* is a bony process extending backward from the pectoral girdle immediately above the base of the pectoral fin. It lies beneath the skin where its outline may be determined by dissecting the skin or probing with a needle.

The retina of catfish are composed of single cones and large rods. Many catfish have a tapetum lucidum which may help enhance photon capture and increase low-light sensitivity. Double cones, though present in most teleosts, are absent from catfish.

The anatomical organisation of the testis in catfish is variable among the families of catfish, but the majority of them present fringed testis: Ictaluridae, Claridae, Auchenipteridae, Doradidae, Pimelodidae, and Pseudopimelodidae. In the testes of some species of Siluriformes, organs and structures such as a spermatogenic cranial region and a secretory caudal region are observed, in addition to the presence of seminal vesicles in the caudal region. The total number of fringes and their length are different in the caudal and cranial portions between species. Fringes of the caudal region may present tubules, in which the lumen is filled by secretion and spermatozoa. Spermatocysts are formed from cytoplasmic extensions of Sertoli cells; the release of spermatozoa is allowed by breaking of the cyst walls.

The occurrence of seminal vesicles, in spite of their interspecific variability in size, gross morphology and function, has not been related to the mode of fertilization. They are typically paired, multi-chambered, and connected with the sperm duct, and have been reported to play a glandular and a storage function. Seminal vesicle secretion may include steroids and steroid glucuronides, with hormonal and pheromonal functions, but it appears to be primarily constituted of mucoproteins, acid mucopolysaccharides, and phospholipids.

Fish ovaries may be of two types: gymnovarian or cystovarian. In the first type, the oocytes are released directly into the coelomic cavity and then eliminated. In the second type, the oocytes are conveyed to the exterior through the oviduct. Many catfish are cystovarian in type, including *Pseudoplatystoma corruscans*, *P. fasciatum*, *Lophiosilurus alexandri*, and *Loricaria lentiginosa*.

### Communication

Sound Production and Interpretation:

Catfish can produce different types of sounds and also have well-developed auditory reception used to discriminate between sounds with different pitches and velocities. They are also able to determine the distance of the sound's origin and in what direction it originated. This is a very important fish communication mechanism, especially during agonistic and distress behaviour. Catfish are able to produce a variety of sounds for communication that can be classified into two groups: drumming sounds and stridulation sounds. The variability in catfish sound signals differs due to a few factors: the mechanism by which the sound is produced, the function of the resulting sound, and physiological differences such as size, sex, and age. In order to create a drumming sound, catfish utilise an indirect vibration mechanism using a swimbladder. In these fishes, sonic muscles insert on the ramus Mulleri, also known as the elastic spring. The sonic muscles pull the elastic spring forward and extend the swimbladder. When the muscles relax, the tension in the spring quickly returns the swimbladder to its original position, which produces the sound.

Catfish also have a sound generating mechanism in their pectoral fins. Many species in the catfish family possess an enhanced first pectoral fin ray, called the spine, which can be moved by large abductor and adductor muscles. The base of the catfishes' spine has a sequence of ridges, and the spine normally slides within a groove on the fish's pelvic girdle during routine movement; but, pressing the ridges on the spine against the pelvic girdle groove creates a series of short pulses. The movement is analogous to a finger moving down the teeth of a comb, and consequently a series of sharp taps is produced.

Sound-generating mechanisms are often different between genders. In some catfishes, pectoral fins are longer in males than in females of similar length, and differences in the characteristic of the sounds produced were also observed. Comparison between families of the same order of catfish demonstrated family and species-specific patterns of vocalization, according to a study by Maria Clara Amorim. During courtship behaviour in three species of *Corydoras* catfishes, all males actively produced stridulation sounds before egg fertilization, and the species' songs were different in pulse number and sound duration.

Sound production in catfish may also be correlated with fighting and alarm calls. According to a study by Kaatz, sounds for disturbance (i.e., alarm) and agonistic behaviour were not significantly different, which suggests that distress sounds can be used to sample variation in agonistic sound production. However, in a comparison of a few different species of tropical catfish, some fish that were put under distress conditions produced a higher intensity of stridulatory sounds than drumming sounds. Differences in the proportion of drumming versus stridulation sounds depend on morphological constraints, such as different sizes of drumming muscles and pectoral spines. Due to these constraints, some fish may not even be able to produce a specific sound. In several different species of catfish, aggressive sound production occurs during cover site defence or during threats from other fish. More specifically, in long-whiskered catfishes drumming sounds are used as a threatening signal and stridulations are used as a defence signal. Kaatz investigated 83 species from 14 families of catfish, and determined that catfishes produce more stridulatory sounds in disturbance situations and more swimbladder sounds in intraspecific conflicts.

### *Catfish as Food*

Catfish have widely been caught and farmed for food for hundreds of years in Africa, Asia, Europe, and North America. Judgments as to the quality and flavour vary, with some food critics considering catfish as being excellent food, while others dismiss them as watery and lacking in flavour. In Central Europe, catfish were often viewed as a delicacy to be enjoyed on feast days and holidays. Migrants from Europe and Africa to the United States brought along this tradition, and in the Southern United States, catfish is an extremely popular food. The most commonly eaten species in the United States are the channel catfish and the blue catfish, both of which are common in the wild and increasingly widely farmed. Farm-raised catfish became such a staple of the diet of the United States that on June 25, 1987, President Ronald Reagan established National Catfish Day to recognise "the value of farm-raised catfish."

Catfish is eaten in a variety of ways. In Europe it is often cooked in similar ways to carp, but in the United States it is popularly crumbed with cornmeal and fried.

In Indonesia, catfish is usually served grilled in street stalls called *warung* and eaten with vegetables and soy sauce; the dish is called *pecel lele*. Catfish can also be eaten with chili sambal as *lele penyet* (minced catfish). (*Lele* is the Indonesian word for catfish.)

In Malaysia catfish, called "ikan keli", is fried with spices or grilled and eaten with tamarind and Thai chillies gravy and also is often eaten with steamed rice.

In Bangladesh and the Indian states of Odisha, West Bengal and Assam catfish (locally known as Magur) is eaten as a favoured delicacy during the monsoons. Catfish, locally known as *thedu or* etta in Malayalam, is very famous in the Indian state Kerala. In the inland ponds in Kerala, 2 varieties of catfish is abundant- Muzhi and Kari while "Etta" is a basically a salt water fish. The smaller, slender Kari is notorious for its ability to sting, and Muzhi is much bigger and easy to catch, especially during Monsoon when this seems to literally walk where very little water is present from the rain water. All the catfish are eaten as curry and their extra-large eggs, especially that of Etta, is fried and is a delicacy. It is also believed that catfish meat helps in blood purification. Catfish curry is consumed in these parts to promote faster recovery to patients suffering from fever or other ailments.

In Hungary catfish is often cooked in paprika sauce (Harcsapaprikás) typical of Hungarian cuisine. It is traditionally served with pasta smothered with curd cheese (túrós csusza).

Catfish is high in Vitamin D. Farm-raised catfish contains low levels of omega-3 fatty acids and a much higher proportion of omega-6 fatty acids.

Vietnamese catfish cannot be legally marketed as catfish in the United States, and is subsequently referred to as *swai* or *basa* Only fish of the family Ictaluridae may be marketed as catfish in the United States.

As catfish lack scales, they are judged not to be kosher and may not be eaten by observant Jews, some Christians who follow the Torah's food restrictions, and observant Muslims of various schools.

### *Aquaculture*

Catfish are easy to farm in warm climates, leading to inexpensive and safe food at local grocers. About 60% of U.S. farm-raised catfish are grown within a 65-mile (100-km) radius of Belzoni, Mississippi. Channel catfish (*Ictalurus punctatus*) supports a $450 million/yr aquaculture industry.

Catfish raised in inland tanks or channels are considered safe for the environment, since their waste and disease should be contained and not spread to the wild.

In Asia, many catfish species are important as food. Several walking catfish (Clariidae) and shark catfish (Pangasiidae) species are heavily cultured in Africa and Asia. Exports of one particular shark catfish species from Vietnam, *Pangasius bocourti*, has met with pressures from the U.S. catfish industry. In 2003, The United States Congress passed a law preventing the imported fish from being labeled as catfish. As a result, the Vietnamese exporters of this fish now label their products sold in the U.S. as "basa fish." Trader Joe's has labeled frozen fillets of Vietnamese *Pangasius hypophthalmus* as "striper."

There is a large and growing ornamental fish trade, with hundreds of species of catfish, such as *Corydoras* and armored suckermouth catfish (often called plecos), being a popular component of many aquaria. Other catfish commonly found in the aquarium trade are banjo catfish, talking catfish, and long-whiskered catfish.

### Catfish as Invasive Species

Representatives of the genus *Ictalurus* have been introduced into European waters in the hope of obtaining a sporting and food resource. However, the European stock of American catfishes has not achieved the dimensions of these fish in their native waters, and have only increased the ecological pressure on native European fauna. Walking catfish have also been introduced in the freshwaters of Florida, with the voracious catfish becoming a major alien pest there. Flathead catfish, *Pylodictis olivaris*, is also a North American pest on Atlantic slope drainages. *Pterygoplichthys* species, released by aquarium fishkeepers, have also established feral populations in many warm waters around the world.

### Dangers to Humans

While the vast majority of catfish are harmless to humans, a few species are known to present some risk. Perhaps the most notorious of these is the candiru, due to the way it is reputed to parasitize the urethra, though there is only one documented case of a candiru attack on a human.

Since 2007, the Goonch catfish has also gained attention following a series of fatal underwater attacks which have been alleged by biologist Jeremy Wade to have been from unusually large goonch.

The Wels catfish has also been reputed to kill humans (especially young children), and while there are no documented cases of fatalities, larger specimens are known to cause serious injuries in rare instances. In addition, other species are reputed to be dangerous to humans as well, but with less definitive evidence. Many catfish species have "stings" (actually non-venomous in most cases) embedded behind their fins; thus precautions must be taken when handling them.

### Taxonomy

The catfishes are a monophyletic group. This is supported by molecular evidence.

Catfish belong to a superorder called the Ostariophysi, which also includes the Cypriniformes, Characiformes, Gonorynchiformes and Gymnotiformes, a superorder characterized by the Weberian apparatus. Some place Gymnotiformes as a sub-order of Siluriformes, however this is not as widely accepted. Currently, the Siluriformes are said to be the sister group to the Gymnotiformes, though this has been debated due to more recent molecular evidence. As of 2007 there are about 36 extant catfish families, and about 3,093 extant species have been described. This makes the catfish order the second or third most diverse vertebrate order; in fact, 1 out of every 20 vertebrate species is a catfish.

The taxonomy of catfishes is quickly changing. In a 2007 and 2008 paper, *Horabagrus*, *Phreatobius*, and *Conorhynchos* were not classified under any current catfish families. There is disagreement on the family status of certain groups; for example, Nelson (2006) lists Auchenoglanididae and Heteropneustidae as separate families, while the All Catfish Species Inventory (ACSI) includes them under other families. Also, FishBase and the Integrated Taxonomic Information System lists Parakysidae as a separate family, while this group is included under Akysidae by both Nelson (2006) and ACSI. Many sources do not list the recently revised family Anchariidae. The family Horabagridae, including *Horabagrus*, *Pseudeutropius*, and *Platytropius*, is also not shown by some authors but presented by others as a true group. Thus, the actual number of families differs between authors. The species count is in constant flux due to taxonomic work as well as description of new species. On the other hand, our understanding of catfishes should increase in the next few years due to work by the ACSI.

The rate of description of new catfishes is at an all-time high. Between 2003 and 2005, over 100 species have been named, a rate three times faster than that of the past century. In June, 2005, researchers named the newest family of catfish, Lacantuniidae, only the third new family of fish distinguished in the last 70 years (others being the coelacanth in 1938 and the megamouth shark in 1983). The new species in Lacantuniidae, *Lacantunia enigmatica*, was found in the Lacantun river in the Mexican state of Chiapas.

According to morphological data, Diplomystidae is usually considered to be the most primitive of catfishes and the sister group to the remaining catfishes, grouped in a clade called Siluroidei. Recent molecular evidence contrasts the prevailing hypothesis, where the suborder Loricarioidei are the sister group to all catfishes, including Diplomystidae (Diplomystoidei) and Siluroidei; though they were not able to reject the past hypothesis, the new hypothesis is not unsupported. Siluroidei was found to be monophyletic without Loricarioid families or Diplomystidae with molecular evidence; morphological evidence is unknown that supports Siluroidei without Loricarioidea.

Below is a list of family relationships by different authors. Lacantuniidae is included in the Sullivan scheme based on recent evidence that places it sister to Claroteidae.

## Aquaculture in China

China, with one-fifth of the world's population, accounts for two-thirds of the worlds reported aquaculture production. Aquaculture is the farming of fish and other aquatic life in enclosures, such as ponds, lakes and tanks, or cages in rivers and coastal waters. China's 2005 reported harvest was 32.4 million tonnes, more than 10 times that of the second-ranked nation, India, which reported 2.8 million tonnes.

China's 2005 reported catch of wild fish, caught in rivers, lakes, and the sea, was 17.1 million tonnes. This means that aquaculture accounts for nearly two-thirds of China's reported total output.

The principal aquaculture-producing regions are close to urban markets in middle and lower Yangtze valley and the Zhu Jiang delta.

### Early History

Aquaculture began about 3500 BC in China with the farming of the common carp. These carp were grown in ponds on silk farms, and were feed silkworm nymphs and faeces. Carp are native to China. They are good to eat, and they are easy to farm since they are prolific breeders, do not eat their young, and grow fast. The original idea that carp could be cultured most likely arose when they were washed into ponds and paddy fields during monsoons. This would lead naturally to the idea of stocking ponds.

In 475 BC, the Chinese politician Fan Li wrote the earliest known treatise on fish farming, *Yang Yu Ching (Treatise on fish breeding)*. The original document is in the British Museum.

During the Tang dynasty (618–907 AD), the farming of common carp was banned because the Chinese word for common carp (É›) sounded like the emperor's family name, *Li* (Ng). Anything that sounded like the emperor's name could not be kept or killed.

The ban had a productive outcome, because it resulted in the development of polyculture, growing multiple species in the same ponds. Different species feed on different foods and occupy different niches in the ponds. In this way, the Chinese were able to simultaneously breed four different species of carp, the mud carp, which are bottom feeders, silver carp and bighead carp, which are midwater feeders, and grass carp which are top feeders. Another development during the Tang dynasty was a fortunate genetic mutation of the domesticated carp, which led to the development of goldfish.

From 1368 AD, the Ming Dynasty encouraged fish farmers to supply the live fish trade, which dominates Chinese fish sales to this day. From 1500 AD, methods of collecting carp fry from rivers and then rearing them in ponds were developed."

### Recent History

The major carp species used traditionally in Chinese aquaculture are the black, grass, silver and bighead carp. In the 1950s, the Pearl River Fishery Research Institute of the Chinese Academy of Fishery Sciences (CAFS) made a technological breakthrough in the induced breeding of these carps, induced by injecting fish pituitary hormones.

In the past, fish culture in China has been a family business, with traditional techniques passed from generation to generation. However, in the late 1960s the Chinese government began a move to the modern induced breeding technologies, which has resulted in a rapid expansion of freshwater aquaculture in China.

From 1978, China's economic policies moved from central planning towards a market economy, opening new markets for aquaculture products. The effect of this, together with further technological advances, has been to move Chinese aquaculture towards industrial scale levels of production. In the 1980s, many species other than carp, such as other species of fish, crustaceans, molluscs and seaweeds, have been brought into production. However, in the late 1990s, CAFS scientists developed a new variant of the common carp called the Jian carp. This succulent fish grows rapidly and has a high feed conversion rate. Over 50% of the total aquaculture production of carp in China has now converted to Jian carp. By 2004, the induced breeding of carps had been so effective that the carp industry amounted to 46 percent of the total aquaculture output.

***Statistics:*** Since 2002, China has been the world largest exporter of fish and fish products. In 2005, exports, including aquatic plants, were valued at US$7.7 billion, with Japan, the United States and the Republic of Korea as the main markets. In 2005, China was sixth largest importer of fish and fish products in the world, with imports totalling US$4.0 billion.

In 2003, the global per capita consumption of fish was estimated at 16.5 kg, with Chinese consumption, based on her reported returns, at 25.8 kg. The common carp is still the number one fish of aquaculture. The annual tonnage of common carp, not to mention the other cyprinids, produced in China exceeds the weight of all other fish, such as trout and salmon, produced by aquaculture worldwide.

Since the 1970s, the reform policies have resulted considerable development of China's aquaculture, both marine and inland. The total used for aquaculture went from 2.86 million hectares in 1979 to 5.68 million hectares in 1996. Over the same time span, production increased from 1.23 million tonnes to 15.31 million tonnes.

In 2005, worldwide aquaculture production including aquatic plants was worth US$78.4 billion. Of this, the Chinese production was worth US$ 39.8 billion. In the same year there were about 12 million fish farmers worldwide. Of these, China reported 4.5 million employed full time in aquaculture.

***Top 10 Species Grown in China in 2005***

| ***Species*** | ***Tonnes*** |
|---|---|
| Japanese kelp | 4 314 000 |
| Grass carp | 3 857 000 |
| Pacific cupped oyster | 3 826 000 |
| Silver carp | 3 525 000 |
| Japanese carpet shell | 2 857 000 |
| Common carp | 2 475 000 |
| Wakame | 2 395 000 |
| Bighead carp | 2 182 000 |
| Crucian carp | 2 083 000 |
| Yesso scallop | 1 036 000 |

***Production, Area and Yield: 2003***

| | ***Total Production(tons)*** | ***Area Used(ha)*** | ***Yield(kg/ha)*** |
|---|---|---|---|
| Overall total | 30,275,795 | 7,103,648 | 4,260 |
| Marine culture | 12,533,061 | 1,532,152 | 8,180 |
| Inland culture | 17,742,734 | 5,571,496 | 3,180 |
| Pond | 12,515,093 | 2,398,740 | 5,220 |

*Contd...*

| | *Total Production(tons)* | *Area Used(ha)* | *Yield(kg/ha)* |
|---|---|---|---|
| Lake | 1,051,930 | 936,262 | 1,120 |
| Reservoirs | 1,841,245 | 1,660,027 | 1,110 |
| Rivers | 738,459 | 382,170 | 1,930 |
| Rice paddies | 1,023,611 | 1,558,042 | 660 |
| Other | 572,396 | 194,297 | 2,950 |

### Inland Aquaculture

In 1979, inland aquaculture occupied 237.8 million hectares and produced 813,000 tonnes. In 1996, they occupied 485.8 million hectares and produced 10.938 million tonnes. In that year, 17 provinces produced 100,000 tonnes from inland aquaculture.

Pond culture is the most common method of inland aquaculture (73.9% in 1996). These ponds are mostly found around the Pearl River basin and along the Yangtze River. They cover seven provinces: Anhui, Guangdong, Hubei, Hunan, Jiangsu, Jiangxi and Shandong. The government has also supported developments in rural areas to get rid of poverty. The sector is significant from a nutrition point of view, because it brings seafood to areas inland away from the sea where consumption of seafood has traditionally been low.

In recent times, China has extended its skills in culturing pond system to open waters such as lakes, rivers, reservoirs and channels, by incorporating cages, nets and pens. Fish farming in paddy fields is also developing. In 1996, paddy fish farming occupied 12.05 million hectares producing 376,800 tonnes. A further 16 million hectares of paddy fields are available for development. Species introduced from other parts of the world are also being farmed, such as rainbow trout, tilapia, paddle fish, toad catfish, silver salmon, river perch, roach and *Collossoma brachypomum*.

Besides fish and crustaceans, turtles (primarily, the Chinese Soft-shelled Turtle *Pelodiscus sinensis*) have been extensively farmed as well since the 1980s and 1990s. Based on a 2002 survey of 684 turtle farms, researchers estimated that these farms had the total herd of more than 300 million animals; they sold over 128 million turtles each year, with the total weight of about 93,000 tons, worth around US$750 million. Since these data are based on less than half of all turtle farms registered with the appropriate regulating agencies (i.e., 684 out of 1,499), it was estimated that the overall herds and production amounts are at least twice as high.

### Marine Aquaculture

Using current culture technologies, much farmed cultivation of marine plants and animals can be applied within the 10 metre isobath in marine environments. There are about 1.33 million hectares of marine cultivable areas in China, including shallow seas, mudflats and bays. Before 1980, less than nine percent of these areas were cultivated, and species were mainly confined to kelp, laver (Porphyra) and mussels. Between 1989 and 1996, areas of cultivated shallow sea were increased from 25,200 to 114,200 hectares, areas of mudflat from 266,800 to 533,100 hectares, and areas of bay from 131,300 to 174,800 hectares. The 1979 production was 415,900 tonnes on 117,000 hectares, and the 1996 production was 4.38 million tonnes on 822,000 hectares.

Since the 1980s, the government has encouraged the introduction of different marine species, including the large shrimp or prawn *Penaeus chinensis*, as well as scallop, mussel, sea bream, abalone, grouper, tilapia and the mud mangrove crab *Scylla serrata*.

In 1989, production of farmed shrimp was 186,000 tonnes, and China was the largest producer in the world. In 1993 viral disease struck, and by 1996 production declined to 89,000 tonnes. This was attributed to inadequate management such as overfeeding and high stock densities.

### Over Reporting

In 2001, the fisheries scientists Reg Watson and Daniel Pauly expressed concerns in a letter to *Nature*, that China was over reporting its catch from wild fisheries in the 1990s. They said that made it appear that the global catch since 1988 was increasing annually by 300,000 tonnes, whereas it was really shrinking annually by 350,000 tonnes. Watson and Pauly suggested this may be related to China policies where state entities that monitor the economy are also tasked with increasing output. Also, until recently, the promotion of Chinese officials was based on production increases from their own areas.

# 5

# Raising Fish for Food in Southeast Asia

Globally aquaculture has been increasing rapidly and already accounts for nearly half of all food fish consumed. For developing countries, which produce 90% of the world's output, aquaculture is a source of protein, employment, income and of foreign exchange. Southeast Asia is an area which has experienced this "blue revolution".

Total aquaculture output in the region increased from less than two million tonnes in 1990 to more than eight million tonnes in 2006. Moreover, the region's pace of expansion has accelerated. Annual average growth rates in output from 2000 to 2006 were more than double those from 1990 to 2000. Already more than a quarter of food fish in Southeast Asia comes from aquaculture.

Aquaculture matters because fish products are important in the diet of much of Southeast Asia. The population generally has a high per capita consumption of fish, and fish are a major source of animal protein in a region where levels of animal protein are below the world average. Output from the capture fisheries has increased but growth rates are slowing. To maintain present levels of per capita consumption of fish in the region, whose average population is projected to grow by 16% by 2015, requires continued expansion of aquaculture.

Over the past years Southeast Asia has become the powerhouse of global fish and shrimp farming and has a clear vision to further expand its aquaculture production. With food safety and environmental protection as key requirements of today's aquaculture, our organic solutions are playing a growing role in the farming of fish and shrimp.

## Southeast Asian Fisheries Development Centre

The Southeast Asian Fisheries Development Centre (SEAFDEC) is an autonomous intergovernmental body established as a regional treaty organisation in 1967 to promote fisheries development in Southeast Asia. SEAFDEC aims specifically to develop the fishery potentials in the region through training, research and information services to improve the food supply by rational utilisation and development of the fisheries resources. Its services cover the broad areas of fishing gear technology, marine engineering, fishing ground surveys and stock assessment, post-harvest technology as well as development and improvement of aquaculture techniques. SEAFDEC is currently made up of 11 Member Countries, namely Brunei Darussalam, Cambodia, Indonesia, Japan, Lao PDR, Malaysia, Myanmar, the Philippines, Singapore, Thailand and Vietnam; and has the Council of Directors, composed of nominees from Member Countries, as policy-making body to provide directives and guidance on activities of the Centre.

The Centre has a Secretariat as its administrative arm, and four technical Departments, namely the Training Department (TD) in Thailand, the Marine Fisheries Research Department (MFRD) in Singapore, the Aquaculture Department (AQD) in the Philippines, and the Marine Fishery Resources Development and Management Department (MFRDMD) in Malaysia.

## Pond Fertilization and Fish Feeds

China has a long history of pond fertilization for fish culture. Farmers adopted the method of manuring to rear fry ages ago. For example, "dacao" (green manure) is used in Guangdong and Guangxi provinces and human excrement (night soil) is used in Jiangxi and Hunan provinces to nurture fry into summer fingerlings. In fingerling-rearing ponds, fertilization is aimed at developing natural food organisms and saving artificial feeds.

Phytoplankton, the elementary producers of the pond, carry out photosynthesis, converting the inorganic materials in the water into the organic nourishment needed for their growth and reproduction. Fertilization supplies the phytoplankton with the materials essential for photosynthesis. As the phytoplankton photosynthesize and reproduce, zooplankton, which feed on phytoplankton, flourish. In turn, the fish, which feed on zooplankton, phytoplankton, and benthos, also flourish. Therefore, the importance of pond fertilization lies in the cultivation and propagation of various food organisms for the cultured fish.

The series of interrelations between predators and prey is called a "food chain." In culturing ponds, fish are the final link: e.g., phytoplankton'!silver carp; phytoplankton'!zooplankton'!bighead; aquatic plants'!grass carp; plankton '!benthos'!black carp. Usually animals use only 5–20 per cent of the energy in both animal and plant feeds. Utilisation of energy is related to the length of the food chain: the shorter the food chain, the higher the rate of energy transfer. In other words, the higher the utilisation rate of energy, the higher the fish production.

The biota of the ponds is in a constant process of growth and decay. Dead organisms are decomposed from complex organic materials into simple inorganic materials by bacteria. These inorganic materials dissolve in water and are utilised by phytoplankton in photosynthesis. Hence, the materials in the pond are in a constant state of circulation mainly through the food chain. This is called "pond material circulation."

### *Varieties of Organic Manure*

Organic manures are mainly farm animal excrement. Generally, the term refers to manures containing organic matter. Today, mainly organic manures are applied to fish ponds in China. The following manures are often used: feces and urine of livestock and poultry, night soil, green manure, compost, arnd silkworm dregs. Only through decomposition by microorganisms is the organic manure converted to nutrients that the plants can absorb.

### *Feces and Urine of Livestock and Poultry*

*Pig manure* — Pig manure includes much organic matter and other nutritional elements such as nitrogen, phosphorus, and potassium and is a fine, complete manure. Pig feces are delicate, containing more nitrogen than other livestock feces (C:N = 14:1), making them more susceptible to rotting. The major portion of pig urine is nitrogen in the form of urea. It decomposes easily.

***Table:*** *Nutritional elements in pig manure.*

| *Item* | *Organic matter (%)* | *Inorganic matter (%)* | | |
|---|---|---|---|---|
| | | *N* | $P_2O_5$ | $K_2O$ |
| Feces | 15 | 0.6 | 0.5 | 0.4 |
| Urine | 2.5 | 0.4 | 0.1 | 0.7 |

The excretory amount of a pig is greatly associated with its body weight and food intake. A 50-kg pig discharges around 10 kg/day or 20 per cent of its body weight. A pig excretes 1000 kg of feces and 1200 kg of urine in the growing period of 8 months from pigling to adult. A pig's daily excretory amount is less than a cow's or a horse's;

however, pigs are advantageous because of their faster growth, shorter fattening period, and suitability for pen culture. Also, pigs are raised on much larger scale, so it is beneficial to collect their manure.

*Cattle manure* — The elements of cattle manure are similar to those of pig manure, but cattle are ruminants and the food stuffs are repeatedly masticated, making the excrement quite delicate. Cattle manure contains less nitrogen than pig manure (C:N = 25:1). Cattle urine contains more nitrogen than pig urine (in the form of hippuric acid, $C_6H_5$ $CONHCH_2$ COOH); therefore, cattle excreta decompose slowly. The average daily excreta is 25 kg/cow, in which the ratio of feces to urine is about 3:2. The total annual amount of excrement for each animal is 9000 kg.

***Table:*** *Nutritional elements in cattle manure.*

| *Item* | *Organic matter (%)* | *Inorganic matter (%)* | | |
|---|---|---|---|---|
| | | *N* | *$P_2O_5$* | *$K_2O$* |
| Feces | 14.0 | 0.3 | 0.2 | 0.1 |
| Urine | 2.3 | 1.0 | 0.1 | 1.4 |

*Poultry manure* — Poultry manures include the feces of chickens, ducks, and geese, and are rich in both organic and inorganic matter.

***Table:*** *Nutritional elements in poultry manure.*

| *Item* | *Organic matter (%)* | *Inorganic matter (%)* | | |
|---|---|---|---|---|
| | | ***N*** | ***$P_2O_5$*** | ***$K_2O$*** |
| Chicken feces | 25.5 | 1.63 | 1.54 | 0.83 |
| Duck feces | 26.2 | 1.14 | 1.44 | 0.62 |
| Goose feces | 23.4 | 0.55 | 0.50 | 0.95 |

Poultry manures rot quickly and their nitrogen is mostly in the form of uric acid, which cannot be absorbed directly by plants. Accordingly, poultry manures are more effective after fermentation. The annual amount of excrement per fowl is as follows: chicken, 5.0–5.7 kg; duck, 7.5–10.0 kg; goose, 12.5–15.0 kg. Although the annual excretory amount of each is comparatively small, the quantity of poultry culture is often great; therefore, the total amount of feces is significant.

### Night Soil

The composition of night soil (human excrement) is greatly dependent on the food consumed. Nitrogen is abundant (C:N = 3:1) and 70–80 per cent of it is in the form of urea. This facilitates easy decomposition.

***Table:*** *Nutritional elements in human excreta.*

| *Item* | *Organic matter (%)* | *Inorganic matter (%)* | | |
|---|---|---|---|---|
| | | ***N*** | ***$P_2O_5$*** | ***$K_2O$*** |
| Feces | 20 | 1.0 | 0.5 | 0.4 |
| Urine | 3 | 0.5 | 0.1 | 0.2 |

On average, an adult excretes 790 kg/year of waste material. This is equivalent to 22 kg/year of $(NH_4)_2SO_4$.

***Table:*** *Yearly excretion of waste by an adult human.*

| *Item* | *Annual amount (kg)* | *Equivalent (kg/year)* | | |
|---|---|---|---|---|
| | | *$(NH_4)_2SO_4$* | *Calcium superphosphate* | *Potassium sulphate* |
| Feces | 90 | 4.5 | 2.25 | 0.7 |
| Urine | 700 | 17.5 | 4.55 | 2.8 |
| Total | 790 | 22.0 | 6.80 | 3.5 |

Night soil to be used as manure must be fermented before application. This is easily done by storing the manure in anaerobic conditions for 2–4 weeks.

The decomposition of human waste produces ammonia. Under airtight conditions, the accumulation of ammonia can sterilize human waste.

Quicklime (1–2 per cent) and formalin (0.1–0.2 per cent) are effective in killing the harmful pathogens in night soil.

### *Silkworm Dregs*

Silkworm dregs are composed of silkworm feces and slough and mulberry residues. They are rich in organic matter: dried dregs are 87 per cent organic matter and 13 per cent nitrogen. They also make good fish feed: 8 kg silkworm dregs can produce 1 kg fish.

### *Green Manure*

All wild grasses and cultivated plants, if used as manure, are called green manures.

These manure rot and decompose easily, providing ideal environments for bacteria propagation. Therefore, they are good for application in fish ponds.

***Table:*** *Nutritional elements in green manures (% wet weight).*

| *Item* | *N* | *$P_2O_5$* | *$K_2O$* |
|---|---|---|---|
| Stems and leaves of broad bean (*Vicia faba*) | 0.55 | 0.12 | 0.45 |
| Rape (*Brassica napus*) | 0.43 | 0.26 | 0.44 |
| Alfalfa (*Medicago falcata*) | 0.54 | 0.14 | 0.40 |
| Wild grass | 0.54 | 0.15 | 0.46 |
| Branyard grass (*Echlnochloa crusgalli*) | 0.35 | 0.05 | 0.28 |
| Alligator weed (*Alternanthera philoxeroides*) | 0.20 | 0.09 | 0.57 |
| Water hyacinth (*Eichhornia crassipes*) | 0.24 | 0.07 | 0.11 |
| Water lettuce (*Pistia stratiotes*) | 0.22 | 0.06 | 0.10 |

### *Compost*

Mixed compost consists of green manure and animal waste. Mixing several manures together may produce a fertilizer that is more suitable to plankton reproduction. The ratio of the constituents depends upon the local sources of manure. Experimental data show the following two mixtures to be suitable for plankton reproduction: (1) green grass — cattle feces — human excreta — lime, 8:8:1:0.17; (2) green grass — cattle feces — lime, 1:1:0.02.

Lime is included in the compost mixture to neutralize the organic acids produced during rotting and decomposition. If these acids were allowed to accumulate, they would inhibit the microorganisms responsible for decomposing the organic matter. There are two methods of making compost; heaping and soaking.

***Heaping Method*** — The manure heap must be made in aerobic conditions. Spread out a layer of green grass, sprinkle some lime on it, add layer of fecal manure, and repeat the procedure. When the compost reaches 1.5–2 m, cover it with 5–6 cm of mud. The ingredients of the compost will rot and decompose. After 3–4 weeks the compost can be used.

***Soaking*** — Dig a pit near the fish ponds and layer in green grass, lime and fecal manure, respectively and then add enough water to soak the compost ensuring there is no leakage. The compost can be removed for use after 10–20 days of fermentation at a temperature of 20–30°C.

## Methods of Organic Manure Application

### *Application of Dacao*

In Guangdong and Guangxi provinces, Dacao is commonly used to fertilize pond water. Dacao consists mostly of composite plants, with some gramineous plants and leguminous plants included. Dacao is applied by heaping it at a corner of the pond and turning the pile once every 2 days. The rotten parts will spread into the water. The roots and stems, which rot slowly, are dredged out of the ponds when the Dacao pile is depleted.

The decomposition of green manure in water consumes a great amount of oxygen. Experimental data show that if 1000 kg of grass is applied to a 1-mu pond with a water depth of 1 m, there will be no available oxygen from the 2nd to the 6th day and all the fish will die. The peak of oxygen consumption because of decomposition is on the 2nd and 3rd days. For this reason, it is appropriate to apply green manures frequently and in small amounts, to frequently add fresh water to the pond, or to use aerators in the pond to guarantee sufficient oxygen for the fish.

### *Application of Night Soil*

In Jiangxi and Hunan provinces, night soil is commonly applied to the fish ponds. Before application, one part night soil is diluted with two parts water. This dilution is then sprayed along the pond dikes once a day. The quantity depends on the fertility of the water and the size of the fish.

### *Application of Livestock Manure*

The application of livestock manure as a base manure is similar to the method for green manure: heap the manures at a corner of the pond or in small piles in shallow water and with a sunny exposure, allow them to decompose and spread gradually into the water. If the manure is used as an additive, it is added in small quantities every 7–10 days.

### *Application of Mixed Compost*

After fermentation, the compost is flushed. The liquid is collected and the residue is removed. This liquid is sprayed into the pond around the dikes. In the case of a large pond, the manure may be loaded on a boat, flushed in batches with pond water, and sprayed evenly over the pond. The manure dregs can be used to fertilize crops. Alternative method is to flip the compost and expose the liquid. The appropriate amount of manure liquid can then be ladled out and spread into the ponds.

The nutrients of the compost are quickly absorbed by phytoplankton. They consume less dissolved oxygen because the organic materials are already decomposed.

## Effects of Manure Application on Natural Food Organisms

The application of organic manure results in the rapid multiplication of bacteria. Bacteria are added with the fertilizer and use the nutrients to reproduce. Also, organic detritus is rich with bacteria, which are an important food source for the lower aquatic animals and filtering fish.

The initial, predominant species of plankton depends closely on the properties of the manure applied. If organic manure is applied, phytoplankton, (*Ochromonas* spp, *Cryptomonas*) and zooplankton

(*Urotrichia* spp.), which are fond of organic materials, will appear first. For inorganic manure, the initial, predominant species will be centric diatoms (*Centomonas* spp. and *Scenedesmus acuminatus*). There is a close relationship between the amount of manure applied and the make-up of the plankton community. Large amounts of manure will lead to the presence of some species of green algae (*Chlorophyta*) and blue algae (*Cyanophyta*); however, small amounts of manure will lead to the presence of *Navicula rostellata* and *Cyclotella stelligera*.

After each manure application, the nutrient content of the water increases, resulting in a planktonic peak. Phytoplankton that are easily digested by silver carp reach a peak after 4 days; those phytoplankton that are not so easily digested attain a climax in 5–10 days. Zooplankton reach a peak in 4–7 days. Protozoans will be the first zooplankton to reach a peak, followed by rotifers, cladocerans, and, finally, copepods. Protozoans multiply by binary fission, increasing the population very rapidly, and, therefore, reaching a peak first. Rotifers usually multiply by parthenogenesis, producing an average of 10–20 eggs during their lifetime.

This process is less productive than binary fission and, therefore, rotifers reach a population peak slightly later than protozoans. Cladocerans also reproduce parthenogenically, but the span between hatching and sexual maturity is longer than that of rotifers; therefore, the cladoceran population peaks later. Copepods take longer time than cladocerans to get mature and its population becomes maximized later. The timing of manure application is crucial when preparing a nursery pond. Ideally, the peak in plankton population should coincide with the feeding demand of the fish fry.

### Varieties of Inorganic Manure

Inorganic manures are also referred to as chemical fertilizers. According to composition, chemical fertilizers can be divided into three groups: nitrogenous, phosphoric, and potash fertilizers. The advantages of inorganic fertilizers are their exact composition, their fast effect, the lack of pollution, their beneficial effect on oxygen content (requiring no decomposition), the small amount required, and their convenient utilisation. However, when chemical fertilizers are applied in ponds, the first link of the food chain is principally phytoplankton, which are not as nutritious to zooplankton as are bacteria. Therefore, the zooplankton population in ponds treated with inorganic manure often lags far behind that in ponds treated with organic manure. Moreover, in most chemical-fertilizer ponds, the predominant phytoplankton is *Chlorophyta*, which is not as nutritious as the predominant phytoplankton in ponds treated with organic manure (*Chrysophyceae, Bacillatiophyceae,* and *Cryptophyceae*). Another disadvantage is that the effect of inorganic fertilizer is rather short and it is difficult to control the water quality. Taking all these factors into account, therefore, the result of chemical-fertilizer application alone is no better than that of organic-fertilizers application.

### Nitrogenous Fertilizers

***Liquid Ammonia*** — Molecular formula: $NH_4OH$ or $NH_3.H_2O$. Nitrogen content: 12–16 per cent. Liquid ammonia is an aqueous solution of ammonia, which is an important product of small-scale nitrogenous fertilizer factories and is easily synthesized at a low cost. Aqueous ammonia is readily volatilized and should not be exposed to the air for a long time.

***Ammonium Sulphate*** — Molecular formula: $(NH_4)_2SO_4$. Nitrogen content: 20–21 per cent. Ammonium sulphate is produced from liquid ammonia directly neutralized with diluted sulphuric acid. When pure, it is a water-soluble white crystal: 75 kg of ammonium sulphate will dissolve in 100 L of water at 20°C. It is easily conserved and applied.

***Urea — Molecular Formula:*** $CO(NH_2)_2$. Nitrogen content: 44–46 per cent. Under high heat and pressure, ammonia and carbon dioxide react to form urea. It is a white crystal with a high solubility in water. However, urea does not ionize when dissolved in water and, therefore, cannot be directly absorbed by plants. It can be utilised by plants only after it has been broken down by urease, excreted from urea-decomposing bacteria, and transformed into ammonium carbonate. This process is temperature dependent in normal ponds. At 20°C, total transformation into ammonium carbonate requires 4–5 days; at 30°C, 2 days.

### Phosphoric Fertilizers

***Calcium Superphosphate*** — Main contents: $Ca(H_2PO_4)_2.H_2O$ with 12–18 per cent $P_2O_5$. Subsidiary contents: $CaSO_4.2H_2O$, about 50 per cent. Calcium superphosphate is usually a white powder, and apt to absorb moisture. It is corrosive and has an acidic odour because it contains some free acids.

## Methods of Inorganic Manure Application

Nitrogen is an essential nutritional element of plants. It is also an essential component of proteins, accelerates the formation of plant chlorophyll, and stimulates photosynthesis. For these reasons, nitrogen content is a decisive factor in phytoplankton production.

Nitrogen is commonly lacking in pond water, so nitrogenous fertilizers should be added. Generally, nitrogenous fertilizer should be used as an additive because of its quick effectiveness. A nitrogenous fertilizer with ammonium must not be mixed with strong alkaline materials e.g. lime; this would result in the volatilization of the ammonium. When using a nitrogenous fertilizer containing ammonium, the toxicity of ammonia must be considered. In aqueous solution, an equilibrium exists between ammonia ($NH_3$) and ammonium ($NH_4^+$).

$$NH_3 + H_2O \rightleftharpoons NH_4^+ + OH^-$$

In an acidic state, the equilibrium shifts to the right and the concentration of ammonium ions increases. In an alkaline state, the equilibrium shifts to the left and the ammonia concentration increases. At a water temperature of 25°C, the percentage of nitrogen as $NH_3$ at various pHs is as follows: ph 6, 0.05 per cent; pH 7, 0.49 per cent; pH 8, 4.7 per cent; pH 9, 32.9 per cent; pH 10, 83.1 per cent; pH 11, 98 per cent.

Ammonia is toxic to fish. It poisons juvenile rainbow trout at 0.3–0.4 mg/L. Chinese carps can tolerate concentrations up to 13 mg/L. Ammonia concentrations below this inhibit growth. The maximum $NH_3$ concentration permitted for fish farming is 0.1 mg/L. Therefore, the amount per application must be strictly controlled. In addition, ponds pH must be closely monitored to avoid applying $NH_3$ in strong alkaline water (e.g., just after pond clearing with lime); liquid ammonia is also alkaline itself. The amount of unionized ammonia increases with increasing water temperature. Therefore, special care is needed when nitrogenous fertilizer in the ammonia form is applied in the summer and the autumn.

The application amount of the nitrogenous fertilizer depends on its nitrogen content. In a pond with an area of 1 mu and a water depth of about 1.5 m, 1.5–2 kg N may be applied as base manure. After this, 0.5 kg N/mu is applied 3 or 4 times monthly. In average, 10 kg N are needed for the whole culture period. For example, if the nitrogen content of ammonium sulphate is about 20 per cent to apply 2 kg of nitrogen to a 1-mu pool as base manure, 10 kg of ammonium sulphate is required. The amount of ammonium sulphate required for a culture period can be calculated in the way: 50 kg.

To apply, make a solution and spread it near the dikes. In the case of liquid ammonia, put the container underwater and open the lid to let the liquid ammonia slowly diffuse out. In this way, volatilization can be avoided.

Most water sources lack phosphorus. Phosphoric fertilizer, besides being utilised by phytoplankton, will also accelerate the reproduction of azotobacteria and complement the nitrogenous fertilizer. The application amount can be calculated based on the phosphoric acid content of the fertilizer. A 0.5–1 kg/mu is used as base manure, the amount used for a culture period is about 5 kg. The method of calculation and application is the same as that for nitrogenous fertilizer.

Potassium is also an essential nutritional element of plants. However, it is usually sufficient in the water and there is no particular need to apply potash fertilizer.

## Fish Feeds

### Significance of Feeding

In addition to fertilizing ponds for the proliferation of natural food organisms, artificial feeds must also be used to meet the demands of various species of fish. Fish feeds are the prime material base of intensive fish culture. Applying artificial feeds in a fish pond can significantly raise per-unit yield. With common carp, for example, output does not exceed 25–30 kg/mu in extensive culture; in intensive culture, however, output will be as high as 200–250 kg/mu. This increase in yield is due to the direct effect of artificial feeding.

Plankton feeder output is also enhanced. The feed is directly consumed by the so-called feed eaters and, in turn, their excreta fertilizers the pond water. This multiplies the natural food organisms of the plankton feeders. In such a culture system, the yield of these species often accounts for one-third of the total fish output.

### Requirements for Different Nutrients

The nutrients that fish require are the same as those required by other animals: proteins, carbohydrates, lipids, vitamins, and minerals. The demand for these elements forms the basis for the preparation and selection of artificial fish feeds.

### Protein

Just like other animals, fish consume protein and break it down into its component amino acids via the enzymes of the digestive system. The amino acids are then absorbed internally and used for normal growth, mending wears, maintenance and reproduction. Protein is used as a source of energy when fats and carbohydrates are depleted;

1 g protein yields 4 Kcal. The dietary protein requirement of farmed fish is generally 25–40 per cent. Terrestrial animals such as chickens, pigs, or cattle usually require 12–17 per cent protein. Because fish are cold-blooded, they require comparatively low-energy feeds. Different fish have different feeding habits and, therefore, different requirements for protein. Carnivorous fish such as rainbow trout and eel demand feeds with high protein contents; herbivorous fish require less protein.

The nutritional value of the feed depends not only on the quantity but also on the quality of the protein, i.e., its amino acid conformation. Amid acids are the elementary units of proteins. Several amino acids are essential to fish growth and development and, therefore, must be present in the feeds. These amino acids are called essential amino acids. The rest of the amino acids, which may or may not exist in the feed, are called dispensable amino acids. They are needed in only small amounts or can be synthesized internally by the fish.

The following 10 amino acids are essential to fish: isoleucine, leucine, lysine, methionine, phenylalanine, threonine, tryptophan, valine, arginine, and histidine. The proportion of these essential amino acids in the feed protein should reflect the amino acid composition of fish protein. Proteins of low nutritional value may be supplemented with the essential amino acids they lack.

### Carbohydrates

Carbohydrates are also defined as polyhydroxy aldehydes or ketones. Through digestion, they are decomposed into monosaccharides, which are absorbed and utilised by the fish. Some monosaccharides are oxidized into carbon dioxide and water: 1 g carbohydrate yields 4 Kcal. Monosaccharides are transported to the liver and muscles and stored as glycogen or are converted into fat which serves as a reserve energy source during food shortages.

Cellulose, a carbohydrate, is the major component of plant cell walls. Among cultivated fish, only a few varieties such as tilapia and milkfish can digest cellulose; even then, the digestibility is low. Cyprinids appear to lack the cellulolytic enzyme and, therefore, cannot digest cellulose. A small amount of cellulose in the fish feed is beneficial, however, because it stimulates the digestive movements of the intestines, promoting the digestion and absorption of other nutrients.

### Lipids

Lipids are also a source of energy; 1 g lipid yields 9 Kcal. During digestion, lipids are broken down into fatty acids and glycerol components, which are absorbed by the fish. Body fats are synthesized from excess fatty acids and glyceryl and are stored in subcutaneous tissues, muscles, spaces between connective tissues, and the abdominal cavity. Lipids tend to deteriorate through oxidation, producing substances that are toxic to fish and destroy vitamin E in fish feeds (e.g., aldehydes and ketones). If fish consume too much deteriorated fish meal and silkworm pupae, which contain oxydized lipids, they will suffer from "thin-back" disease, muscular atrophy, and weight loss, and show a high mortality rate.

### *Vitamins*

Vitamins are essential organic compounds required by fish in trace quantities. There are many varieties of vitamins with various physiological functions; however, all of them are essential if fish are to grow and develop normally. Some vitamins participate in the metabolic process: vitamin $B_1$, carbohydrate metabolism; vitamin $B_6$, protein metabolism; vitamin C, protein synthesis; vitamin D, calcium and phosphorus metabolism (formation of the skeleton).

It is difficult to determine the exact amounts of the various vitamins required by the fish. In Chinese fish culture, fresh feeds are added to prevent vitamin deficiencies. For example, if pelleted feeds are used in grass carp farming, green grasses will be supplemented to ensure a complete supply of vitamins.

### *Minerals*

Minerals are essential elements in many aspects of fish metabolism and structure. For example, phosphorus and calcium are important components of the skeleton. A deficiency of either substance will result in deformity. Both fish feeds and pond water serve as a source of minerals. Fish absorb calciferous and phosphorous salts and chlorine and sodium ions through their gill and skin.

Minerals enhance the utilisation of carbohydrates, accelerate the growth of certain tissues (e.g., skeleton and muscles), and enhance the fish's appetite; therefore, minerals must be included in the preparation of fish feeds. Additives such as bone powder and table salt in fish feeds fulfill the requirements of the pond fish.

## Varieties of Fish Feeds

### *Plant Feeds*

*Grain* — Soybean, wheat, and maize are commonly used grain feeds. Soybean, which is a nutritional food, contains at least 38 per cent crude protein (Table 3.9) and is rich in essential amino acids. Soybeans are usually ground up into "bean milk" for feeding to fry. About 5–7 kg of soybeans can supply the milk required to raise 10,000 fish from fry to summerling. In nurturing fry, some bean milk particles are directly consumed by the fry but most become fertilizer for the proliferation of plankton. For grass carp brooders, wheat and rice sprouts are often supplied. These grains are rich in vitamins, especially vitamin E, which is beneficial to gonad development. For granulated feeds, wheat powder is often used as a binder.

*Oil Cakes* — Cakes are by-products of oil plants after oil pressing. Bean cakes, peanut cakes, cottonseed cakes, and rapeseed cakes are often used in fish farming. This type of feed is rich in crude protein (30–40 per cent). If cakes are used to feed fry, they must be broken into pieces, soaked, and ground up into a milk. About 150–200 kg of cake is needed to nurture 10,000 fingerlings with fsa body length greater than 10 cm.

***Table:*** *Food conversion ratio (FCRs) of several common feeds for various species of fish.*

| *Feed* | *Species* | *FCR* |
|---|---|---|
| Snails | Black carp | 40 |
| | Common carp | 50 |
| Clams (*Corbicula*) | Black carp | 80 |
| | Black carp | 60 |
| Soybean cake | Black carp | 3 |
| | Common carp | 3.5 |
| Rapeseed cake | Black carp | 4.0 |
| | Common carp | 4.5 |
| Peanut cake | Black carp | 3 |
| | Common carp | 4 |
| Cotton seed cake | Grass carp | 6 |
| Rye | Black carp | 4 |
| | Grass carp | 3 |
| Barley | Black carp | 4 |
| | Grass carp | 3 |
| Silkworm | | |
| Fresh pupae | Black carp | 3.5 |
| Dry pupae | Black carp | 1.5 |
| | Common carp | 2 |

*Contd...*

| ***Feed*** | ***Species*** | ***FCR*** |
|---|---|---|
| Aquatic grass | Grass carp | 90 |
| | Bream | 100 |
| Terrestrial grass | Grass carp | 40 |
| | Bream | 45 |
| Duck weeds | Grass carp | 50 |
| | Bream | 50 |
| Rye grass | Grass carp | 25 |
| | Bream | 30 |
| Sudan grass | Grass carp | 40 |
| Pumpkin vine | Grass carp | 35 |
| Wheat bran | Common carp | 4 |
| | Tilapia | 4 |
| Rice bran | Common carp | 3.5 |
| | Tilapia | 3.5 |
| Bean dregs | Grass carp | 25 |
| | Bream | 25 |

Cottonseed cakes are commonly used in carp culture in the USA and the USSR. This is not so common in China. However, China is a cotton-producing country and the use of cottonseed cake as a fish feed shows great potential. There is a little gossypol in cottonseed cakes, which is detrimental to livestock, but harmless to fish.

*Wheat bran* — Wheat bran is a by-product of the rice- and wheat-processing industry and is rich in vitamin B, apart from crude proteins, fats, and carbohydrates. It is an important ingredient of compound feeds.

***Table:*** *Nutritional elements of various plant feeds (%).*

| | ***Moisture*** | ***Crude protein*** | ***Crude fat*** | ***Crude fibre*** | ***Non-nitrogenous extract*** |
|---|---|---|---|---|---|
| Soybean cake | 11.3 | 39.1 | 7.1 | 4.5 | 32.0 |
| Peanut cake | 11.3 | 38.4 | 8.2 | 5.8 | 29.5 |
| Cottonseed cake | 9.3 | 35.0 | 6.0 | 10.1 | 30.3 |
| Rapeseed cake | 11.0 | 31.0 | 6.7 | 8.2 | 31.1 |
| Rice bran | 11.8 | 10.8 | 12.0 | 8.2 | 47.0 |
| Wheat bran | 13.1 | 10.9 | 3.7 | 8.9 | 55.3 |
| Soybean | 11.2 | 38.1 | 13.1 | 4.1 | 27.5 |
| Barley | 14.5 | 10.0 | 2.0 | 4.0 | 69.0 |
| Maize | 12.0 | 8.5 | 4.3 | 1.3 | 71.0 |

*Green fodders* — Green fodders include aquatic plants and terrestrial plants and are mainly used as feed for grass carp and breams and, sometimes, for common carp, crucian carp, and tilapia. The main aquatic plants used are *Wolffia arrhiza, Lemna minor, Vallisneria spiralis, Potamogeton malainus, Potamogeton maackianus, Hydrilla verticillata, Eichhornia crassipes, Pistia stratiotes,* and *Alternanthera philoxeroides.* The main terrestrial plants used are *Echinochloa crusgalli, Pennisetum alopecuroides, Lolium perenne, Sorghum sudanense, Pennisetum purpurlum* of the grass family; *Lactuca tenticulata* of the composite family; and various leaves and vines from melon and vegetable crops.

Water lettuce (*Pistia stratiotes*), water hyacinth (*Eichhornia crassipes*), and alligator weed (*Alternanthera philoxeroides*) must be minced or fermented before they are given to the fish. To ferment one of these aquatic plants, 100 kg is mixed with 3–4 kg rice bran and 0.5 kg yeast; the container is then sealed and stored for 2 days at 26°C. Alligator weed sometimes contains saponin, which is toxic to fish. If this is processed by adding a little table salt (2–5 per cent concentration), the toxicity is eliminated and the feed becomes palatable. When these aquatic plants are mashed into a paste with a high-speed masher, they are appropriate feeds for fry culture. The fry swallow the mesophyll cells in the paste, which are similar in size to zooplankton and phytoplankton. The material in the paste serves as manure for the reproduction of plankton and other natural foods.

***Table:*** *Composition of some green fodders (%).*

| ***Fodder*** | ***Moisture*** | ***Crude protein*** | ***Crude fat*** | ***Crude fibre*** | ***Non-nitrogenous extract*** |
|---|---|---|---|---|---|
| *Wolffia arrhiza* | 96.02 | 1.25 | 0.41 | 0.38 | 1.52 |
| *Lemna minor* | 95.80 | 1.43 | 0.38 | 0.42 | 1.23 |
| *Vallisneria spiralis* | 96.77 | 0.61 | 0.09 | 0.66 | 1.17 |
| *Potamogeton malainus* | 89.41 | 2.11 | 0.17 | 2.17 | 4.89 |
| *Potamogeton maackianus* | 87.18 | 2.16 | 0.46 | 3.11 | 5.65 |
| *Leersia japonica* | 74.61 | 3.72 | 1.27 | 7.50 | 10.70 |
| *Sorghum sudanense* | 82.83 | 1.78 | 0.69 | 4.75 | 8.66 |
| *Lactuca tenticulata* | 88.95 | 3.02 | 0.95 | 1.60 | 4.20 |
| *Pistia stratiotes* | 91.90 | 1.20 | 0.40 | 1.80 | 2.90 |
| *Eichhornia crassipes* | 94.90 | 1.00 | 0.20 | 0.90 | 1.80 |
| *Alternanthera philoxeroides* | 77.50 | 3.22 | 0.80 | 2.62 | 11.62 |
| *Lolium perenne* | 85.35 | 4.17 | 0.59 | 3.54 | 4.43 |

Green fodders contain mostly water and cellulose. However, they also contain the principal nutrients, i.e., fat, protein and carbohydrate, and are rich in vitamins. Green fodders are the principal feed for grass carp and wuchang fish and serves as supplemental feed for other cultivated fish.

### Animal Feeds

Animal feeds have a higher nutritional value than plant feeds because they give a more complete supply of nutrients and are richer in proteins and essential amino acids. Among the cultivated fish species in China, most are herbivorous or omnivorous; the only carnivorous species is black carp.

Common animal feeds include fish meal, trash fish, silkworm pupae, fresh-water shellfish (e.g., *Viviparus quadratus* and *Corbicula fluminea*), kitchen waste, and earthworms. *Viviparus quadratus* lives in rivers and lakes with a high fecundity and feeds on epiphytic algae.

It has a meat rate of 22–25 per cent and a food-conversion ratio of 40 for black carp. *Corbicula fluminea* lives in the clay bottom of rivers and lakes and is collected with *Viviparus quadratus*. It has a meat rate of 13 per cent and a food conversion ratio of 60 for black carp. Both species are commonly fed to black carp.

***Table:*** *Composition (%) of common animal feeds.*

| | | *Moisture* | *Crude protein* | *Crude fat* | *Non-nitrogenous extract* |
|---|---|---|---|---|---|
| Fish meal | | 10.0 | 59.0 | 9.8 | 0.4 |
| Silkworm pupae | (Fresh) | | 17.1 | 9.2 | |
| | (Dry) | 7.3 | 56.9 | 24.9 | 4.0 |
| Snail | | 75.8 | 14.1 | 0.4 | |
| *Corbicula* | | 85.0 | 5.3 | 2.0 | 7.0 |

In Japan and China, silkworm pupae are commonly used as fish feed. Fresh pupae are more effective but harder to preserve than dry pupae, which are rich in fats; however, these fats deteriorate easily through oxidization. Furthermore, the fish fed with silkworm pupae have off-flavour; therefore, pupae feeding is stopped 2–3 weeks before harvesting.

Pig blood is used as a binding agent for pelleted feeds. These feeds are effectively used in rainbow trout culture.

### Formulated Feeds

*Advantages* — Formulated feeds are composed of several materials in various proportions. In fish farming, formulated feeds have the following advantages:

- The ingredients of formulated feeds can complement one another and raise the food utilisation rate.
- Proteins can supplement one another improving the essential amino acid conformation of the feed and raising the protein utilisation rate.
- Food sources can be broadened by mixing feeds disliked by the fish with other preferred feeds.
- By adding a binding agent to produce pelleted feeds, the solution of nutrients in water is diminished and wastage is reduced.
- Drugs may be mixed into the feeds (indicated feeds) to control fish diseases.
- Formulated feeds are convenient to transport and preserve; they are suited to automatic feeding, which can lead toward the mechanization of fish farming.

*Preparation* — The feeding habits and nutritional requirements of the fish and the nutritional content of the food must be considered when making a nutritionally balanced feed. To achieve complete nutrition, many feedstuffs should be combined. The economics of formulating the feed should be considered; a good selection of cheap materials is preferable. Different formulas should reflect the different nutritional requirements of various species at different developmental stages; e.g., more protein for black carp than for grass carp; more protein for finger-lings than for adults. A moderate amount of binder should be added to ensure a high water stability of the feeds.

Some formulated pellet feeds have been tried on black carp because of a lack of snails and *Corbicula.*

- "320" straw powder pellet feed: "320" straw powder is made by fermenting "320" Basidiomycetes: straw powder, 50 per cent; bean cake powder, 10 per cent; fish meal powder, 5 per cent; barley flour, 15 per cent; wheat bran 10 per cent; rapeseed cake 10 per cent. The crude protein content is 20.14 per cent and the recommended daily feeding rate is 3–5 per cent of body weight. The food conversion ratio for 2-year-old black carp in monoculture is 2.
- Bean cake powder, 10 per cent; pupae, 10 per cent; barley flour, 30 per cent; rapeseed cake, 50 per cent; bone powder, 2 per cent; table salt, 1 per cent. The crude protein content is 28.1 per cent and the carbohydrate content is 38.9 per cent. The food conversion ratio for 2-year-old black carp in monoculture is 2.4.

- Bean cake powder, 25 per cent; fish meal, 4 per cent; barley flour, 16 per cent; rapeseed cake, 24 per cent; wheat bran, 26.5 per cent; mineral mixture, 1.5 per cent; plant oil, 3 per cent.

Formulated feeds have also been used with grass carp.

- Straw powder, 25 per cent; sesame stem powder, 25 per cent; bean cake powder, 25 per cent; ricebran, 25 per cent; waste flour (Binder), 10 per cent. The crude protein content is 23.35 per cent. The food conversion ratio is 2.8.
- Straw powder, 70 per cent; bean cake powder, 15 per cent; ricebran, 10 per cent; waste flour, 5 per cent; bone powder, 1 per cent; table salt, 0.5 per cent. The crude protein content is 15.07 per cent. The food conversion factor is 4.4.
- "320" green grass powder pellet feed: green grasspowder, 40 per cent; rapeseed cake, 20 per cent; fish meal, 5 per cent; bean cake, 15 per cent; silkworm pupae, 5 per cent; barley flour, 15 per cent. The crude protein content is 22.6 per cent. The recommended daily feeding rate is 3.5 per cent of body weight. The food conversion rate for grass carp and wuchang fish is 2.5–3.

If only pellet feeds are applied, the muscle, intestine and liver of grass carp accumulate a high amount of fat. This can have a pathological effect on the liver and, consequently, adversely affect growth. When green grass is also provided (5 days of pellet feeds followed by 2 days of grass application), this situation improves: no pathological change is observed.

Nevertheless, every substance demanded by the fish cannot be included in a specific feed, no matter how complete the feed is. Therefore, it is important to also supply fresh, natural fodders (e.g., green grass for carp and snails for black carp) to promote the digestion of the artificial diet.

# 6

# Biochemistry of Fish Oils

Fish oil is oil derived from the tissues of oily fish. Fish oils contain the omega-3 fatty acids eicosapentaenoic acid (EPA), and docosahexaenoic acid (DHA), precursors of certain eicosanoids that are known to reduce inflammation throughout the body, and have other health benefits.

Fish do not actually produce omega-3 fatty acids, but instead accumulate them by consuming either microalgae or prey fish that have accumulated omega-3 fatty acids, together with a high quantity of antioxidants such as iodide and selenium, from microalgae, where these antioxidants are able to protect the fragile polyunsaturated lipids from peroxidation. Fatty predatory fish like sharks, swordfish, tilefish, and albacore tuna may be high in omega-3 fatty acids, but due to their position at the top of the food chain, these species can also accumulate toxic substances through biomagnification. For this reason, the U.S. Food and Drug Administration recommends limiting consumption of certain (predatory) fish species (e.g. albacore tuna, shark, king mackerel, tilefish and swordfish) due to high levels of toxic contaminants such as mercury, dioxin, PCBs and chlordane. Fish oil is used as a component in aquaculture feed. More than 50 percent of the world's fish oil used in aquaculture feed is fed to farmed salmon.

Marine and freshwater fish oil vary in contents of arachidonic acid, EPA and DHA. The various species range from lean to fatty and their oil content in the tissues has been shown to vary from 0.7–15.5%. They also differ in their effects on organ lipids. Studies have revealed that there is no relation between total fish intake or estimated omega"3 fatty acid intake from all fish and serum omega"3 fatty acid concentrations. Only fatty fish intake, particularly salmonid, and estimated EPA + DHA intake from fatty fish has been observed to be significantly associated with increase in serum EPA + DHA.

The omega-3 fatty acids in fish oil are thought to be beneficial in treating hypertriglyceridemia, and possibly beneficial in preventing heart disease. Fish oil and omega-3 fatty acids have been studied in a wide variety of other conditions, such as clinical depression, anxiety, cancer, and macular degeneration, although benefit in these conditions remains to be proven.

## Background

The most widely available dietary source of EPA and DHA is cold water oily fish, such as salmon, herring, mackerel, anchovies, and sardines. Oils from these fish have a profile of around seven times as much omega-3 oils as omega-6 oils. Other oily fish, such as tuna, also contain omega-3 in somewhat lesser amounts. Although fish is a dietary source of omega-3, fish do not synthesize them; they obtain them from the algae (microalgae in particular) or plankton in their diets.

***Table:*** *Grams of omega-3 fatty acids per 3oz (85g) serving of popular fish.*

| *Common Name* | *Grams* |
|---|---|
| Herring, sardines | 1.3–2 |
| Spanish mackerel, Atlantic, Pacific | 1.1–1.7 |
| Salmon | 1.1–1.9 |
| Halibut | 0.60–1.12 |
| Tuna | 0.21–1.1 |
| Swordfish | 0.97 |
| Greenshell/lipped mussels | 0.95 |
| Tilefish | 0.9 |
| Tuna (canned, light) | 0.17–0.24 |
| Pollock | 0.45 |
| Cod | 0.15–0.24 |
| Catfish | 0.22–0.3 |
| Flounder | 0.48 |
| Grouper | 0.23 |
| Mahi mahi | 0.13 |
| Orange roughy | 0.028 |
| Red snapper | 0.29 |

*Contd...*

| Common Name | Grams |
|---|---|
| Shark | 0.83 |
| King mackerel | 0.36 |
| Hoki (blue grenadier) | 0.41 |
| Gemfish | 0.40 |
| Blue eye cod | 0.31 |
| Sydney rock oysters | 0.30 |
| Tuna, canned | 0.23 |
| Snapper | 0.22 |
| Barramundi, saltwater | 0.100 |
| Giant tiger prawn | 0.100 |

For comparison, here are the omega-3 levels in some common non-fish foods.

***Table:*** *Grams of omega-3 fatty acids per 3oz (85g) serving of common non-fish foods.*

| *Name* | *Grams* |
|---|---|
| Eggs, large regular | 0.109 |
| Lean red meat | 0.031 |
| Turkey | 0.030 |
| Cereals, rice, pasta, etc. | 0.00 |
| Fruit | 0.00 |
| Milk regular | 0.00 |
| Regular bread | 0.00 |
| Vegetables | 0.00 |

## *Production*

In 2005, fish oil production declined in all main producing countries with the exception of Iceland. The 2005 production estimate is about 570,000 tonnes in the five main exporting countries (Peru, Denmark, Chile, Iceland and Norway), a 12% decline from the 650,000 tonnes produced in 2004.. Peru continues to be the main fish oil producer worldwide, with about one fourth of total fish oil production.

## Health Benefits

### *Cancer*

Several studies report possible anti-cancer effects of *n*"3 fatty acids found in fish oil (particularly breast, colon and prostate cancer).

Among n-3 fatty acids (omega-3), neither long-chain nor short-chain forms were consistently associated with reduced breast cancer risk. High levels of docosahexaenoic acid, however, the most abundant n-3 polyunsaturated fatty acid (omega-3) in erythrocyte membranes, were associated with a reduced risk of breast cancer.

A recent study of 35,000 middle-aged women found that the women who took fish oil supplements had a 32% lower risk of breast cancer, although the authors stress the result is preliminary and falls short of establishing a causal relationship. Omega-3 fatty acids reduced prostate cancer growth, slowed histopathological progression, and increased survival in genetically engineered mice. However the effects of fish oil consumption by humans on prostate cancer is not conclusive. There is a decreased risk with higher blood levels of DPA, but an increased risk of more aggressive prostate cancer with higher blood levels of combined EPA and DHA.

### *Cardiovascular*

The American Heart Association recommends the consumption of 1g of fish oil daily, preferably by eating fish, for patients with coronary heart disease although pregnant and nursing women are advised to avoid eating fish with high potential for mercury contaminants including mackerel, shark, or swordfish. Note that optimal dosage relates to body weight. The US National Institutes of Health lists three conditions for which fish oil and other omega-3 sources are most highly recommended: hypertriglyceridemia, secondary cardiovascular disease prevention and high blood pressure. It then lists 27 other conditions for which there is less evidence. It also lists possible safety concerns: "Intake of 3 grams per day or greater of omega-3 fatty acids may increase the risk of bleeding, although there is little evidence of significant bleeding risk at lower doses. Very large intakes of fish oil/omega-3 fatty acids may increase the risk of hemorrhagic (bleeding) stroke."

There is also some evidence that fish oil may have a beneficial effect on some forms of cardiac dysrhythmia.

A 2008 meta-study by the *Canadian Medical Association Journal* found fish oil supplementation did not demonstrate any preventative benefit to cardiac patients with ventricular arrhythmias. A 2012 meta-analysis published in the Journal of the American Medical Association, covering 20 studies and 68680 patients, found that fish oil supplementation did not reduce the chance of death, cardiac death, heart attack or stroke.

### *Hypertension*

There have been some human trials that have concluded that consuming omega-3 fatty acids slightly reduces blood pressure (DHA could be more effective than EPA). It is important to note that because omega-3 fatty acids can increase the risk of bleeding, a qualified healthcare provider should be consulted before supplementing with fish oil.

### *Mental Health*

Studies published in 2004 and 2009 have suggested that the *n-3* EPA may reduce the risk of depression and suicide. One study compared blood samples of 100 suicide-attempt patients and to those of controls and found that levels of Eicosapentaenoic acid were significantly lower in the washed red blood cells of the suicide-attempt patients. A small American trial in 2009 suggested that E-EPA, as monotherapy, might treat major depressive disorder but failed to achieve statistical significance.

Studies were conducted on prisoners in England where the inmates were fed seafood which contains omega-3 fatty acids. The higher consumption of these fatty acids corresponded with a drop in the assault rates. Another Finnish study found that prisoners who were convicted of violence had lower levels of omega–3 fatty acids than prisoners convicted of nonviolent offences. It was suggested that these kinds of fatty acids are responsible for the neuronal growth of the frontal cortex of the brain which, it is further alleged, is the seat of personal behaviour.

A study from the Orygen Research Centre in Melbourne suggests that omega-3 fatty acids could also help delay or prevent the onset of schizophrenia. The researchers enlisted 81 'high risk' young people aged 13 to 24 who had previously suffered brief hallucinations or delusions and gave half of them capsules of fish oil while the other half received placebo. One year on, only three percent of those on fish oil had developed schizophrenia compared to 28 percent from those on placebo. A study conducted at Sheffield University in England reported positive results with fish oil on patients suffering from schizophrenia. Participants of the study had previously taken anti-psychotic prescription drugs that were no longer effective. After taking fish oil supplements, participants in the study experienced progress compared to others who were given a placebo.

The largest controlled study to date found no cognitive benefit after two years in the elderly.

## Alzheimer's Disease

According to a study from Louisiana State University in September 2005, Docosahexaenoic acid, an omega-3 fatty acid often found in fish oil, may help protect the brain from cognitive problems associated with Alzheimer's disease.

A Cochrane meta-analysis published in June 2012 found no significant protective effect for cognitive decline for those aged 60 and over and who started taking fatty acids after this age. A co-author of the study said to *Time*, "Our analysis suggests that there is currently no evidence that omega-3 fatty acid supplements provide a benefit for memory or concentration in later life".

### *Lupus*

In a study conducted in Northern Ireland, lupus disease activity, especially in the skin and joints, was significantly reduced in patients who received fish oil supplements at both 12-week and 24-week follow-up periods versus patients who received placebo. There were also changes in the blood platelets of the patients who took the fish oil supplements, with an increase in proteins that are considered anti-inflammatory and a decrease in proteins that promote inflammation; these changes were not evident in the group that took placebo. The fish oil group showed an increase in flow-mediated dilation, which the researchers took as a sign that the omega-3 oils were helping the cells in the blood vessel walls to remain healthy.

## Parkinson's Disease

A study examining whether omega-3 exerts neuroprotective action in Parkinson's disease found that it did exhibit a protective effect in mice. The scientists exposed mice to either a control or a high omega-3 diet from two to twelve months of age and then treated them with a neurotoxin commonly used as an experimental model for Parkinson's. The scientists found that high doses of omega-3 given to the experimental group prevented the neurotoxin-induced decrease of dopamine that ordinarily occurs. Since Parkinson's is a disease caused by disruption of the dopamine system, this protective effect exhibited could show promise for future research in the prevention of Parkinson's disease.

### *Depression*

Evidence regarding the efficacy of fish oil supplements as a treatment for depression is inconclusive. Whereas several methodologically rigorous studies have reported statistically significant positive effects in the treatment of depressed patients, other studies have found effects to be insignificant.

In 1999 a team of researchers lead by the Harvard psychiatrist Andrew Stoll published a preliminary placebo-controlled double blind trial which found Omega 3 fatty acids "improved the short-term course of illness" of bipolar disorder. He credits Donald O. Rudin for pioneering this view in 1981.

A 2003 double-blind placebo-controlled study published in the journal *European Neuropsychopharmacology* found that among 28 patients with major depressive disorder, "patients in the omega-3 PUFA group had a significantly decreased score on the 21-item Hamilton Rating Scale for Depression than those in the placebo group." Another study in the *American Journal of Psychiatry* reported that the addition of fish oil supplements to regular maintenance anti-depression therapy conferred "highly significant" benefits by the third week of the trial.

A 2005 randomized double-blind placebo-controlled study conducted under the auspices of the New Zealand Institute for Crop and Food Research found "no evidence that fish oil improved mood when compared to placebo, despite an increase in circulating ù-3 polyunsaturated fatty acids." Another study published in October 2007 found that fish oil supplements conferred no additional benefits beyond those conferred by standard treatment. However, both of these studies used omega-3 primary consisting of DHA, not EPA.

A 2008 Cochrane systematic review found that limited data is available. In the one eligible study, omega-3s were an effective adjunctive therapy for depressed but not manic symptoms in bipolar disorder. The authors found an "acute need" for more randomised controlled trials.

A 2009 metastudy found that patients taking omega-3 supplements with a higher EPA:DHA ratio experienced less depressive symptoms. The studies provided evidence that EPA may be more efficacious than DHA in treating depression. However, this metastudy concluded that due to the identified limitations of the included studies, larger, randomized trials are needed to confirm these findings.

In a 2011 meta-analysis of PubMed articles about fish oil and depression from 1965-2010, researchers found that "nearly all of the treatment efficacy observed in the published literature may be attributable to publication bias."

### Psoriasis

Diets supplemented with cod liver oil have shown beneficial effects on psoriasis.

### Pregnancy

Omega-3 polyunsaturated fatty acids (commonly found in fish oil) protect against fetal brain injury and promotes fetal and infant brain health. Some studies reported better psycho motor development at 30 months of age in infants whose mothers received fish oil supplements for the first four months of lactation. In addition, five-year-old children whose mothers received modest algae based docosahexaenoic acid supplementation for the first 4 months of breastfeeding performed better on a test of sustained attention. This suggests that docosahexaenoic acid intake during early infancy confers long-term benefits on specific aspects of neurodevelopment.

Docosahexaenoic acid supplementation has also been found to be essential for early visual development of the baby. However, the standard western diet is severely deficient in these critical nutrients. This omega-3 dietary deficiency, a nutrient found in fish oil, is compounded by the fact that pregnant women become depleted in omega-3s, since the fetus uses omega-3s for its nervous system development. Omega-3s are also used after birth if they are provided in breast milk.

In addition, provision of fish oil during pregnancy may reduce an infant's sensitization to common food allergens and reduce the prevalence and severity of certain skin diseases in the first year of life. This effect may persist until adolescence with a reduction in prevalence and/or severity of eczema, hay fever and asthma.

Omega-3 fatty acid supplementation is also beneficial to the mother. It has been shown to prevent pre-term labour and delivery. It is recommended that women who are breastfeeding consume fish oil at least twice a week, although the American Heart Association recommends pregnant and nursing women are to avoiding eating fish with high potential for mercury contaminants including mackerel, shark, or swordfish.

## Supplement Quality and Concerns

Fish oil is now one of the most popular dietary supplements on the market, with sales reaching $976 million in 2009. However, neither the FDA nor any other federal or state agency routinely tests fish or marine oil supplements for quality prior to sale. Problems of quality have been identified in periodic tests by independent researchers of marketed supplements containing fish oil and other marine oils. These problems include contamination, inaccurate listing of EPA and DHA levels, spoilage and formulation issues.

### *Contamination*

Fish can accumulate toxins such as mercury, dioxins, and polychlorinated biphenyls (PCBs), and spoiled fish oil may produce peroxides. There appears to be little risk of contamination by microorganisms, proteins, lysophospholipids, cholesterol, and trans-fats.

### *Mercury*

While a serving of fish may contain anywhere from 10 to 1,000 ppb of mercury, fish oil supplements have not been found to contain similar mercury levels. Reasons for this are 1) smaller fish are typically used in making fish oil supplements and they tend to be lower on the food chain and contain less mercury; 2) mercury binds to protein (such as in fish meat) and not to oil; and 3) mercury may be reduced or removed during the processing of fish oil. (most fish oils are distilled)

### *Dioxins and PCBs*

Dioxins and PCBs may be carcinogenic at low levels of exposure over time. These substances are identified and measured in one of two categories, dioxin-like PCBs and total PCBs. While the U.S. FDA has not set a limit for PCBs in supplements, the Global Organisation for EPA and DHA (GOED) has established a guideline allowing for no more than 3 picograms of dioxin-like PCBs per gram of fish oil.

In 2012, samples from 35 fish oil supplements were tested for PCBs. Trace amounts of PCBs were found in all samples, and two samples exceeded the GOED‘s limit. Although trace amounts of PCBs contribute to overall PCB exposure.

### *Spoilage*

Peroxides can be produced when fish oil spoils. A study commissioned by the government of Norway concluded there would be some health concern related to the regular consumption of oxidized (rancid) fish/marine oils, particularly in regards to the gastrointestinal tract, but there is not enough data to determine the risk.

The amount of spoilage and contamination in a supplement depends on the raw materials and processes of extraction, refining, concentration, encapsulation, storage and transportation.

## EPA and DHA Content

Some countries recommend a combined daily intake of 300 – 500 mg while some countries like the US do not have a recommendation for EPA and DHA. The American Heart Association recommends 250–500 mg/day of EPA and DHA. In the United States the FDA recommends not exceeding 3 grams per day of EPA and DHA omega-3 fatty acids, with no more than 2 grams per day from a dietary supplement.

According to independent laboratory tests, the concentrations of EPA and DHA in supplements can vary from between 8 to 80% fish oil content. The concentration depends on the source of the omega-3s, how the oil is processed, and the amounts of other ingredients included in the supplement. A 2012 report claims 4 of 35 fish oil supplements it covered contained less EPA or DHA than was claimed on the label, and 3 of 35 contained more.

## Formulation

Fish oil supplements are available as liquids, capsules, and tablets. Some pills are enteric-coated to help prevent indigestion or "fish burps," however; enteric-coated products have the potential to release ingredients too early or late in the digestive process. Fish oils are best tolerated when taken with meals, and, if possible, should be taken in equally divided doses throughout the day.

## Storage

Fish oil should be stored properly to prevent rancidity.

## Dangers

***Reduced Blood Clotting:*** Decreased clotting caused by fish oil can be dangerous if taking blood-thinning medications such as aspirin, warfarin or clopidogrel. Those with bleeding disorders such as hemophilia may find their condition exacerbated. According to the University of Maryland Medical Centre, fish oil should not be taken with any blood-thinning medications unless under the supervision of a physician.

## Maximum Intake

The FDA says it is safe to take up to 3000 mg of omega-3 per day. (This is not the same as 3000 mg of fish oil. A 1000 mg pill typically has only 300 mg of omega-3; 10 such pills would equal 3000 mg of omega-3.) Dyerberg studied healthy Greenland Eskimos and found an average intake of 5700 mg of omega-3 EPA per day.

### *Vitamins*

The liver and liver products (such as cod liver oil) of fish and many animals (such as seals and whales) contain omega-3, but also the active form of vitamin A. At high levels, this form of the vitamin can be dangerous (Hypervitaminosis A).

### *Toxic Pollutants*

Consumers of oily fish should be aware of the potential presence of heavy metals and fat-soluble pollutants like PCBs and dioxins, which are known to accumulate up the food chain. After extensive review, researchers from Harvard's School of Public Health in the *Journal of the American Medical Association* (2006) reported that the benefits of fish intake generally far outweigh the potential risks.

Fish oil supplements came under scrutiny in 2006, when the Food Standards Agency in the UK and the Food Safety Authority of Ireland reported PCB levels that exceeded the European maximum limits in several fish oil brands, which required temporary withdrawal of these brands. To address the concern over contaminated fish oil supplements, the International Fish Oil Standards (IFOS) Program, a third-party testing and accreditation program for fish oil products, was created by Nutrasource Diagnostics Inc. in Guelph, Ontario, Canada.

A March 2010 lawsuit filed by a California environmental group claimed that eight brands of fish oil supplements contained excessive levels of PCB's, including CVS/pharmacy, Nature Made, Rite Aid, GNC, Solgar, Twinlab, Now Health, Omega Protein and Pharmavite. The majority of these products were either cod liver or shark liver oils. Those participating in the lawsuit claim that because the liver is the major filtering and detoxifying organ, PCB content may be higher in liver-based oils than in fish oil produced from the processing of whole fish.

An analysis based on data from the Norwegian Women and Cancer Study (NOWAC) with regards to the dangers of persistent organic pollutants (POPs) in cod liver came to the conclusion that "in Norwegian women, fish liver consumption was not associated with an increased cancer risk in breast, uterus, or colon. In contrast, a decreased risk for total cancer was found."

However, a report by the Harvard Medical School studied five popular brands of fish oil, including Nordic Ultimate, Kirkland and CVS. They found that the brands had "negligible amounts of mercury, suggesting either that mercury is removed during the manufacturing of purified fish oil or that the fish sources used in these commercial preparations are relatively mercury-free." Microalgae oil is a vegetarian alternative to fish oil. Supplements produced from microalgae oil provide a balance of omega-3 fatty acids similar to fish oil, with a lower risk of pollutant exposure.

## Tilapia

Tilapia is the common name for nearly a hundred species of cichlid fish from the tilapiine cichlid tribe. Tilapia are mainly freshwater fish, inhabiting shallow streams, ponds, rivers and lakes, and less commonly found living in brackish water.

***Figure:*** *Nile tilapia,* Oreochromis niloticus

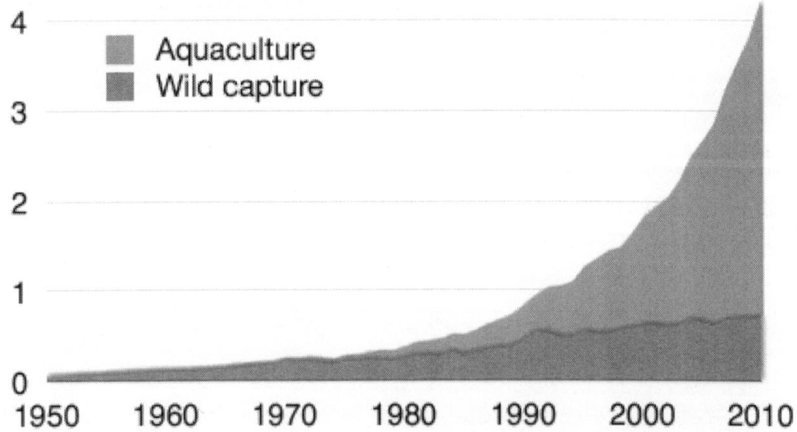


***Figure:*** *Global harvest of tilapia in million tonnes as reported by the FAO, 1950–2009*

Historically, they have been of major importance in artisan fishing in Africa and the Levant, and are of increasing importance in aquaculture and aquaponics. Tilapia can become problematic invasive species in new warm-water habitats, whether deliberately or accidentally introduced, but generally not in temperate climates due to their inability to survive in cooler waters below about 21 °C (70 °F).

### History

Tilapia were one of the three main types of fish caught in Biblical times from the Sea of Galilee. At that time were called *musht*, or commonly now even "St. Peter's fish". The name "St. Peter's fish" comes from the story in the Gospel of Matthew about the apostle Peter catching a fish that carried a coin in its mouth, though the passage does not name the fish. While the name also applies to *Zeus faber*, a marine fish not found in the area, a few tilapia species (*Sarotherodon galilaeus galilaeus* and others) are found in the Sea of Galilee, where the author of the Gospel of Matthew accounts the event took place. These species have been the target of small-scale artisanal fisheries in the area for thousands of years.

The common name tilapia is based on the name of the cichlid genus *Tilapia*, which is itself a latinisation of *thiape*, the Tswana word for "fish". Scottish zoologist Andrew Smith named the genus in 1840.

### Characteristics

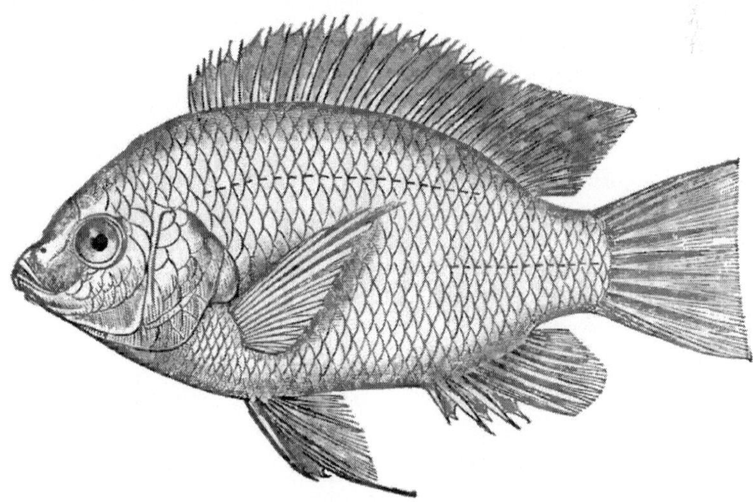

***Figure:*** *Nile tilapia,* Oreochromis niloticus

Tilapia typically have laterally compressed, deep bodies. Like other cichlids, their lower pharyngeal bones are fused into a single tooth-bearing structure. A complex set of muscles allows the upper and lower pharyngeal bones to be used as a second set of jaws for processing food (cf. morays), allowing a division of labour between the "true jaws" (mandibles) and the "pharyngeal jaws". This means they are efficient feeders that can capture and process a wide variety of food items. Their mouths are protrusible, usually bordered with wide and often swollen lips. The jaws have conical teeth. Typically tilapia have a long dorsal fin, and a lateral line which often breaks towards the end of the dorsal fin, and starts again two or three rows of scales.

### *Species*

Tilapia as a common name has been applied to various cichlids from three distinct genera: *Oreochromis*, *Sarotherodon* and *Tilapia*. The members of the other two genera used to belong to the genus *Tilapia* but have since been split off into their own genera. However, particular species within are still commonly called "tilapia" regardless of the change in their actual taxonomic nomenclature.

The delimitation of these genera among each other and to other tilapiines requires more research; mtDNA sequences are confounded because at least among the species of any one genus, there is frequent hybridization. The species remaining in *Tilapia* in particular still seem to be a paraphyletic assemblage.

## Threatened Species

### *Exotic and Invasive Species*

Tilapia has been used as biological controls for certain aquatic plant problems. It has a preference for a floating aquatic plant, duckweed (*Lemna* sp.) but also consume some filamentous algae. In Kenya tilapia were introduced to control mosquitoes which were causing malaria, because they consume mosquito larvae, consequently reducing the numbers of adult female mosquitoes, the vector of the disease (Petr 2000). These benefits are, however, frequently outweighed by the negative aspects of tilapia as an invasive species.

Tilapia are unable to survive in temperate climates because they require warm water. The pure strain of the blue tilapia, *Oreochromis aureus*, has the greatest cold tolerance and dies at 45 °F (7 °C), while all other species of tilapia will die at a range of 52 to 62 °F (11 to 17 °C). As a result, they cannot invade temperate habitats and disrupt native ecologies in temperate zones; however, they have spread widely beyond their points of introduction in many fresh and brackish tropical and subtropical habitats, often disrupting native species significantly. Because of this, tilapia are on the IUCN's 100 of the World's Worst Alien Invasive Species list. In the United States, tilapia are found in much of Florida, Texas and a few other isolated areas, such as power plant discharge zones. Many state fish and wildlife agencies in the United States, Australia, South Africa and elsewhere consider them invasive species.

### *Aquarium Species*

Larger tilapia species are generally poor community aquarium fish because they eat plants, dig up the bottom, and fight with other fish. However, the larger species are often raised in aquariums as a food source, because they grow rapidly and tolerate high stocking densities and poor water quality.

Smaller West African species, such as *Tilapia joka* and species from the crater lakes of Cameroon, are more popular. In specialised cichlid aquaria, tilapia can be mixed successfully with nonterritorial cichlids, armored catfish, tinfoil barbs, garpike and other robust and dangerous fish. Some species, including *Tilapia buttikoferi*, *Tilapia rendalli*, *Tilapia mariae*, *Tilapia joka* and the brackish-water *Sarotherodon melanotheron melanotheron*, have attractive patterns and are quite decorative.

### *Commercial Species*

The tilapiines of North Africa are the most important commercial cichlids. Fast-growing, tolerant of stocking density, and adaptable, tilapiine species have been introduced and farmed extensively in many parts of Asia and are increasingly common aquaculture targets elsewhere.

## Fish in Afghanistan

Afghanistan is a landlocked country, which covers an area of 652,225 km$^2$, nearly 75% of which is mountainous. The average elevation is 1300 m. The climate varies sharply between highlands and lowlands. It is sub-polar in the mountainous northeast with dry, cold winters, with temperatures falling to -26°C or lower in the Hindu Kush range. South of the highlands lies an arid, virtually uninhabited southwestern plateau. There are three great river basins: the Amu-Darya (Oxus), which forms the boundary with Tajikistan, Uzbekistan and Turkmenistan in the north; the Kabul in the northeast, which enters Pakistan in the east where it has a confluence with the Indus; and the Helmand in the southwest, which ends in Iran in a desert lake immediately after crossing the border in the southwest.

The source of surface water in all rivers is precipitation, and consequent snow melt, over the central mountain ranges extending from the Pamir mountain knot at the western termination of the Karakoram, the Hindu Kush and its outliers, and the ranges of Hazarajat.

Maximum water flow is in the spring and early summer and minimum flow is in late summer to winter over much of the country. Many rivers dry up along sections of their course or are reduced to isolated pools during the minimum-flow period. This natural condition is aggravated by water abstraction for irrigation and other purposes,, and rivers tend to disappear before reaching their principal river or lake (Coad, 1981).

In the Pamir and Nurestan areas of the northeast, melting glaciers feed the rivers in July and August, but with the advancing freezing temperatures the flow rate greatly diminishes. Rivers along the northeast border with Pakistan are affected by the monsoon and have maximum flows twice a year: July to September and January to April. There are few freshwater lakes in Afghanistan, the largest being those of Sistan which lie mostly in Iran but are hydrographically part of Afghanistan.

Major perennial rivers are the Amu Darya, Qonduz (=Kunduz), Kowcheh (=Kokcha), Band-e Amir, Kabul, Lowgar (=Logar), Panjsher, Laghman, Konar (=Kunar), Sorkh Ab, Helmand, Arghandab, Hari Rud and Morghab (=Murgab).

The following notes deal only with the major rivers and their major tributaries. Their names, and the names of smaller rivers not mentioned here, will also appear where appropriate in Table 1, which lists the individual fish species and their occurrence.

The Amu Darya River has its sources in the Pamirs, and it ends in the Aral Sea. The lower 1300 km of a total of 2500 km lies wholly outside Afghanistan. Coad included in his list of fish species those which occur in the lower Amu Darya as they may penetrate upriver either naturally or by being stocked there. He did not include fish of the Zarafshan River and its tributaries Kara Darya and Ak Darya, although Zarafshan is now connected with Amu Darya through canals. In its upper reaches the Amu Darya is known as Vakhan River

(Vakhsh), then as Panj River (Pyandzh), when it receives the Pamir River. In Afghanistan the Panj is called the Amu Darya when it is joined by the Kowkcheh (=Kokcha) River. However, in Tajikistan, the name of Amu Darya starts from the entry of the Vakhsh. The Qonduz River enters the Amu Darya near its junction with the Vakhsh River.

The Murgab River (or Morghab) has its source in the western Hindu Kush, flowing west and then north to the Afghanistan-Turkmenistan border, crossing into Turkmenistan, where it is lost in the sands of the Kara Kum Desert.

The Hari Rud River, which starts in the centre of Afghanistan, flows directly west and eventually enters Turkmenistan where it is called Tedzhen. It also ends in the sands of the Kara Kum Desert.

The Helmand River has its source not far away from the source of the Kabul River. It flows southwest for about 1300 km before it empties into the Sistan lakes. The river with its tributaries drains about 40% of Afghanistan and has the largest drainage basin. Only the Helmand River is perennial. It has been dammed in several places.

The Kabul River has its source in the mountains west of Kabul and flows east to join the Indus River north of Attock. The river is dammed in several places. It receives a number of tributaries, such as the Panjsher, Laghman, Konar, Lowgar and Sorkh Rud. The Swat and Khiali rivers enter the Kabul River in Pakistan.

The list of fishes as given in Table is an abbreviated version of a more detailed account of fish of Afghanistan by Coad. Species marked * have not been reported from Afghanistan but occur in adjacent or contiguous drainage basins. Species marked # have been introduced into Afghanistan. The species *Pungitius platygaster* was left out from the list as it occurs only in the Aral Sea and Amu Darya delta.

***Table:*** *List of fish species found in Afghanistan*

| ***Family/Species*** | ***Localities*** |
|---|---|
| Family: Acipenseridae | |
| Acipenser nudiventris | Panj River |
| Pseudoscaphirhynchus hermanni | Amu Darya at Termez |
| Pseudoscaphirhynchus kaufmanni | Panj River |
| Family: Esocidae | |
| *Esox lucius | Amu Darya at Pitnyak (lower reaches of Amu Darya) |

*Contd...*

| ***Family/Species*** | ***Localities*** |
|---|---|
| Family: Salmonidae | |
| Oncorhynchus mykiss | Stocked in 1966 into the Salang and Panjsher |
| Salmo trutta | Also known as S.t.aralensis. Upper Amu Darya; Panj; Bamian |
| Family: Cyprinidae | |
| *Abramis brama | In lakes along the Amu Darya up to above Turtkul |
| *Abramis sapa | Amu Darya up to Pitnyak |
| Alburnoides bipunctatus | Murgab River; upper Amu Darya |
| Alburnoides taeniatus | Amu Darya above Termez; Qonduz; Khanabad |
| Amblypharyngodon mola | Kabul River |
| Aspidoparia jaya | Kabul River |
| *Aspidoparia morar | SE Iran; Swat R. (Pakistan); |
| Aspiolucius esocinus | Amu Darya and lower reaches of its tributaries in Tajikistan; lower Panj River |
| Aspius aspius | Amu Darya up to Kafirnigan River |
| Barbus brachycephalus | Amu Darya up to Fayzabadkala on the Panj |
| Barbus capito | Amu Darya up to Fayzabadkala on the Panj; Qonduz River at Qonduz; Andarab River; |
| Barilius vagra | Rivulet at Khost - Chamkani River drainage; Kabul River; Zhob River; Wana Toi, a tributary of the Gumal River in Pakistan |
| Capoeta capoeta | Helmand; Harirud; Tedzhen; Murgab; Qonduz, upper Amu Darya; Panj |
| *Capoeta fusca | Iranian drainage of the Namakzar |
| Capoetobrama kuschakewitschi | Amu Darya to the Panj; Surkhan; Kafirnigan; Qonduz; Khanabad; |
| #Carassius auratus | Kabul River; Pishin Lora drainage; lower Amu Darya drainage |
| *Chalcalburnus chalcoides | Amu Darya up to Kushkantau |

*Contd...*

| ***Family/Species*** | ***Localities*** |
|---|---|
| Cirrhinus burnesiana | At Jalalabad? |
| Cirrhinus reba | Kabul River |
| Crossocheilus latius | Kurram River (Chamkani R. in Afghanistan) in Pakistan; several other rivers in Pakistan; |
| #Ctenopharyngodon idella | Kabul River drainage |
| Cyprinion watsoni | River at Kushk in NW Afghanistan; Nushki and Quetta in Pakistan; Khost; Sistan; Zhob; Kurram and Wana Toi, both in Pakistan |
| Cyprinus carpio | Amu Darya up to Panj; Murgab; Tedzhen, Lake Gusar; Qonduz, Khanabad |
| Danio devario | Kabul River |
| Esomus danricus | Kabul River |
| *Garra gotyla | Zhob River drainage |
| Garra rossica | Tedzhen; Murgab; Koshk; Shila; Pishin Lora; Helmand; Sistan; Wana Toi (Pakistan) |
| *Garra rufa | eastern Khorasan (Iran) |
| Gobio gobio | Amu Darya at Termez; Kafirnigan; Tedzhen; Murgab; Koshk |
| #Hemiculter leucisculus, | |
| Qonduz; Khanabad | |
| *Hemigarra elegans | Sistan |
| #Hypophthalmichthys molitrix | Kabul River drainage |
| Labeo angra | Kabul River |
| Labeo dero | Kabul River |
| Labeo diplostomus | Kabul River |
| Labeo dyocheilus | Kabul River |
| Labeo gonius | Kabul River |
| Labeo pangusia | Kabul River |
| *Leuciscus idus | Amu Darya up to Pitnyak |
| Leuciscus latus | Haridut; Murgab; Tedzhen |
| *Leuciscus lehmanni | Surkhan Darya; Kafirnigan |
| *Leuciscus leuciscus | Kafirnigan |

*Contd...*

| ***Family/Species*** | ***Localities*** |
|---|---|
| Pelecus cultratus | Amu Darya up to Panj |
| #Pseudorasbora parva | Khanabad |
| Ptychobarbus conirostris | Afghanistan (?) |
| Puntius conchonicus | Kabul River; |
| Puntius sarana | Kabul River |
| Puntius sophore | Kabul River |
| *Punius ticto | Lower Swat River (Pakistan) |
| Rutilus rutilus | Amu Darya up to Pitnyak; Khanabad |
| Salmostoma bacaila | Kabul River |
| *Salmostoma punjabensis | Swat River drainage (Pakistan) |
| *Scardinius erythrophthalmus | Amu Darya up to Pitnyak |
| Schizocypris brucei | Sistan; in Pakistan: Wana Toi; Zhob; Kurram |
| Schizocypris ladigesi | Chamkali drainage; Khost; Ali Khel |
| Schizocypris stoliczkae | Upper Amu Darya in the Pamirs, upper Helmand |
| *Schizothorax anjac | Sistan |
| Schizothorax barbatus | Kabul River |
| Schizothorax chrysochlora | Kabul River; Chamkani drainage; Lowgar |
| Schizothorax edeniana | Kabul River |
| Schizothorax esocinus | Kabul and Helmand drainages; Chitral drainage |
| Schizothorax gobioides | Bamian |
| Schizothorax intermedius | Upper Amu Darya; Panj; Gunt; Bartang; Tanyas Pamir; Indus and Helmand River basins |
| Schizothorax labiatus | Panjsher; Konar; Kabul; in Pakistan Zhob; Chitral and Swat River drainages |
| Schizothorax microcephalus | Panjah (entering the Oxus) |
| Schizothorax pelzmani | Murgab; Tedzhen; Iranian drainages of Tedzhen and Germab rivers |
| Schizothorax plagiostomus | Pich; Salang; Panjsher; Paghman; Dasht-e- Navar; Farrakhollum; in Pakistan Kurram |

*Contd...*

| ***Family/Species*** | ***Localities*** |
|---|---|
| *Schizothorax schumacheri | Sistan |
| Schizothorax zarudnyi | Harut (Adraskan) |
| Tor putitora | Kabul River and its drainage |
| *Tor zhobensis | Zhob River basin (Pakistan) |
| Family: Cobitidae | |
| Noemacheilus (Triplophysa) akhtari | Helmand at Farakhollum |
| Noemacheilus (Schistura) alepidotus | Swat River drainage and the Ghowr Band (=Chorband) River in the Kabul River basin |
| Neomacheilus (Triplophysa) | Amu Darya from Chardzhou to Aivadzh amudarjensis |
| Noemacheilus (Schistura) | Kajaki in Helmand River drainage baluchiorum Sistan |
| Noemacheilus (Paracobitis) | Sistan and the Helmand River drainage boutanensis |
| Noemacheilus (Triplophysa) brahui | Pakistan: Nushki and Pishin |
| *Noemacheilus (Triplophysa) choprai | Pakistan: Chitral and Swat River drainages |
| *Noemacheilus (Schistura) corica | Pakistan: Bannu - drainage of the Kurram |
| Noemacheilus (Paracobitis) cristatus | Murgab; Qual el-Chabrak; Harirud drainage; Helmand drainage; Pakistan: Zhob drainage |
| *Noemacheilus (Deuterophysa) dorsalis | Amu Darya basin in the mountains; Kafirnigan River |
| Noemacheilus (Triplophysa) farwelli | Helmand River |
| Noemacheilus (Paracobitis) ghazniensis | Ghazni River |
| Noemacheilus (Triplophysa) griffithi | Sistan basin; Arghandab; Kabul River and drainage; Chorband; Salang; Paghman Pakistan: Swat |
| Noemacheilus (Schistura) kessleri | Ghazni River; Janichel (?); in Pakistan - Pishin Lora drainage; Nushki; Pishin; Iran - in SE |
| Noemacheilus (Triplophysa) kullmanni | Dasht-e-Navar |
| *Noemacheilus (Oreias) | Dyushambinka River in Kafirnigan |

*Contd...*

| ***Family/Species*** | ***Localities*** |
|---|---|
| kuschakewitschi | River basin |
| Noemacheilus (Paracobitis) | Tedzhen and Murgab drainages; Khanabad |
| longicauda | River |
| Noemacheilus (Paracobitis) | Tedzhen, Murgab, Amu Darya and |
| malapterurus | Panj rivers; Helmand River in Sistan |
| *Noemacheilus (Schistura) naseeri | Pakistan: Swat River |
| Noemacheilus (Orthias) oxianus | Amu Darya up to Panj River |
| *Noemacheilus (Schistura) | Pakistan: Zhob River drainage |
| pakistanicus | |
| Noemacheilus | (Schistura) prashari |
| Noemacheilus (Paracobitis) | Helmand River drainage |
| rhadineus Helmand; Tedzhen; | |
| Murgab; Pakistan: Zhob River | |
| drainage | |
| Noemacheilus (Schistura) | River in the Kabul River basin: Iran: |
| sargadensis | Pech Baluchistan; Turkmenistan: east of Ashgabat |
| *Noemacheilus (Triplophysa) | Helmand catchment; Pakistan: |
| stenurus | Chitral and Swat rivers |
| Noemacheilus (Triplophysa) | Upper Amu Darya; Gunt; Tanymas; |
| stoliczkae | upper Helmand; Indus River mountain basin; Bamian; |
| Noemacheilus (Triplophysa) tenuis | Upper Amu Darya drainage in the Pamirs; Koshk; Gunt; Sistan |
| Sabanajewia aurata | Tedzhen; Murgab; Amu Darya basin to the Pamirs |
| Family: Bagridae | |
| Mystus seenghala | Afghanistan |
| Mystus tengara | Aghanistan |
| Rita rita | Kabul River; Khyber Pass |
| Family: Siluridae | |
| Ompok bimaculatus | Afghanistan |
| Ompok canio | Kabul River |
| Ompok pabda | Kabul River; Pakistan: Wana Toi |

*Contd...*

| ***Family/Species*** | ***Localities*** |
|---|---|
| Silurus glanis | Amu Darya; Qonduz; Murgab |
| Wallago attu | Pakistan: Pishin Lora drainage |
| Family: Schilbeidae | |
| *Clupisoma naziri | Pakistan: Khiali River - a tributary of the Kabul River in Pakistan |
| | Family: Sisoridae |
| Glyptosternum akhtari | Bamian River |
| Glyptosternum reticulatum | Upper Amu Darya; Bamian; Kabul River drainage; Panjsher; Salang; Paghman; Surchab; Chitral valley; Andarab |
| *Glyptothorax cavia | Pakistan: Khiali; Swat River drainage |
| Glyptothorax jalalensis | Kabul River |
| *Glyptothorax naziri | Pakistan: Zhob River drainage |
| *Glyptothorax punjabensis | Pakistan: Khiali River |
| *Glyptothorax stocki | Pakistan: Swat River |
| | Family: Poeciliidae |
| #Gambusia affinis | Sistan; Tedzhen; Murgab; Amu Darya basin; Khanabad; Kabul. Pakistan: Pishin Lora |
| Family: Percidae | |
| *Gymnocephalus cernua | Amu Darya |
| *Perca fluviatilis | Amu Darya |
| *Stizostedium lucioperca | Amu Darya |
| Family: Gobiidae | |
| #Rhinogobius similis | Khanabad |
| Family: Channidae | |
| Channa gachua | Kabul River drainage |
| Channa punctatus | Kabul River drainage |
| Family: Mastacembelidae | |
| *Mastacembelus armatus | Pakistan: Zhob River |

With regards to coldwater fish, Coad (1981) made the following notes. The upper reaches of the Kabul basin are dominated by a variety of snow trout (Schizothoracini) and cobitid species which are adapted to cold, fast-flowing mountain streams. The fish fauna of the upper Amu Darya is impoverished in comparison with that of the lower parts of the drainage basin. Certain species found in the upper Amu Darya are also found in the upper reaches of adjacent drainages. For example *Glyptosternum reticulatum* is also found in the Kabul system and the Tarim basin. The Helmand River is the largest of the three major drainage basins of Afghanistan and has the least diverse ichthyofauna in terms of number of species. *Noemacheilus stoliczkae* and *Schizothorax intermedius* are found in all three major drainages.

Afghanistan rivers and streams contain a mixture of Oriental and Palaearctic species, of northern and southern species and of high and low altitude-adapted species. The fauna is divided between Oriental and Palaearctic species. The fauna is dominated by Cyprinidae (56.9%), Cobitidae (24.5%) and to a lesser extent by Siluriformes (11.8%).

### Capture Fisheries and Aquaculture

Fisheries activities in rivers and streams of Afghanistan have been very limited, and information on the number of fishermen, fish species captured, yields and total catch does not exist. The FAO Yearbook on Fishery Statistics has been publishing estimates of catches, rising from 800 t in 1986 to 1300 t in 1995 as compared to the estimate of 100 t for 1963. The true situation may be considerably different from these estimates, as no concrete data have been submitted by Afghanistan for at least the last 10 years. It is recognised that fish does not contribute much to the economy of the country and therefore is not paid the same attention as other animal resources.

In 1967 a trout fish hatchery was established at Qargha Dam, about 15 km from Kabul. The dam, constructed across the River Paghman, created a water reservoir of about 50 km$^2$. The hatchery was supplied with water from this reservoir. In the 1970s it was producing about 30,000 trout fingerlings, which were stocked into the Qargha Reservoir and the rivers Panjsher, Bamian, Salang and Sarde. The stocking of the reservoir was done largely for licensed sport fishing. In the 1970s a second trout hatchery was located near the town of Paghman, west of Kabul.

In 1987, assistance was provided by the UNDP/FAO to rehabilitate the Qargha Fish Farm near Kabul. During 1988-89 supplies of spring water were restored, egg incubators repaired and fitted with new egg trays, and the hatchery brought back into production. The intention was to produce fish to market size in floating cages moored in the adjacent Qargha Reservoir. Concrete raceways next to the farm were also brought back into usable condition. Rainbow trout were grown from eyed eggs imported from Denmark in 1988, and by 1989 six tons of fingerlings were produced. The deteriorating security situation in 1989 interfered with a successful completion of the project, with much of the trout production being stolen or sold underweight. Only 3 tons were sold, against a target of 10 tons. Nevertheless, the project demonstrated the technical feasibility of culturing rainbow trout at Qargha fish farm.

At the same time a warmwater fish farm was located alongside Darunta, 150 km east of Kabul. This hatchery was completed in 1966 with the assistance of China, and China also provided technical assistance until 1972. Darunta fish farm was producing fingerlings of four carp species (grass, silver, common, and bighead). The fingerlings were stocked into Darunta Reservoir, and resulted in the production of 144.2 t of fish over the period of 1967-1973, with 30 t captured in 1973 (El Zarka). At that time there were 41 fishermen harvesting the reservoir fish. Management of fish stocks in the reservoir faced several problems, such as escape of fish during floods over the flood controlling gates, and the presence of dense aquatic vegetation. By 1973 a decline in catches was also observed.

Nevertheless, it was believed that reservoirs would be important for the future development of inland fisheries. At that time, apart from Darunta, four other reservoirs were situated in not too great a distance from Kabul: Neghlo, Soroby, Arghandab and Kajaki. Kajaki Reservoir was considered to have more favourable conditions for fish production than Arghandab Reservoir, which was considered too oligotrophic, and also subject to drastic drawdown. In 1992 Darunta Dam was seriously damaged in the war.

## A Fish Suq in the UAE Desert

Until the late 1960's, Al Ain was an isolated date palm oasis with a small resident population controlled by the semi-nomadic bedouin of the Beni Yas confederation. The leaders of these tribes had gained control of the oasis by steadily purchasing date palm gardens. The Beni Yas had their main seat of power in the coastal settlement at Abu Dhabi, west of Al Ain, and were the most important tribe in a group of oases in the Liwa area to the south west of Al Ain. Adjacent to the five villages comprising Al Ain are four Omani villages forming the Buraimi part of the oasis. In distance terms the Batinah coast of Oman is as near to Al Ain as the Gulf coast. There are strong cultural ties between the two communities although the strength of tribal custom has ensured that they now belong to separate countries.

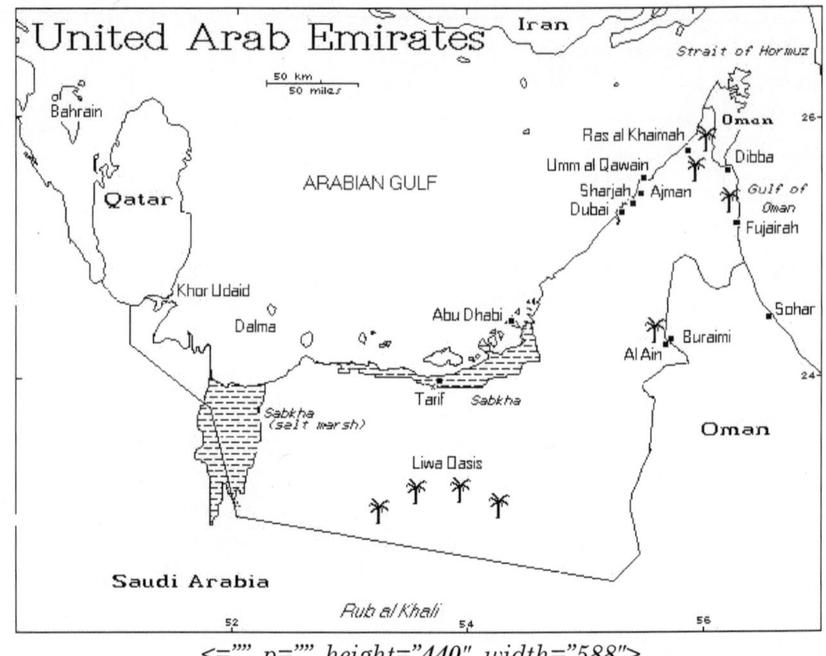


*<="" p="" height="440" width="588">*

The Al Ain fish market is housed in a number of sheds arranged around the open vegetable and food market in the heart of the city. The whole market derives its popular name, suq as samak, from the fish sales. The fish are brought overnight from the coastal landing areas by truck in large ice chests. The trucks are driven into the sheds so that the fish can be heaped on benches for sale as the morning wears on and the stock diminishes. One is confronted by mounds of multi-coloured fish backed up by fishmongers wielding sharp knives gutting two kilo trevally specimens whilst vying for your attention for the next sale. The fish is invariably fresh and the customers are knowledgeable and discerning.

The choice is large and spectacular, heaps of silvery sardines, broomtail wrasse in exquisite colours, vicious barracuda and cutlass fish with bared fangs, lines of svelte tuna, a tangle of half-beaks, half a dozen varieties of grouper with bulging eyes, bowls of swimming crabs and prawns, rather evil looking sea catfish with poisonous spines, and a plethora of trevally and bream.  Individual fish weigh from a few grams, for instance anchovies, to kingfish and amberjack weighing 15-20 Kg. each. The appendix records the technical details of this visual treat and also records local fish names gathered from the other Gulf countries.

The market is well patronised by all nationalities with the exception of Europeans and North Americans. They are perhaps too accustomed to the supermarket culture to cope with the vagaries of an open market and all its questionable characteristics; for instance the need to know about the different fish varieties; bargaining skills and the hygiene aspect.

### Fish Butchery

The term butchery was chosen with care as there is minimal finesse displayed when it comes to cutting up the fish sold to a customer. However the majority of fish sold in the suq are offered for sale intact. In the case of large fish such as tuna, fresh sharks and kingfish, they are usually sliced transversely into steaks. Some fish seem to be prepared as a matter of course, shaeri (emperors) are frequently displayed scaled, gutted, fins trimmed and the head removed. If the vendor cleans a fish for a local he will offer to cut it up and this offer is usually taken up. The fish is then butchered into chunks with no respect for bone structure and must be an alarming prospect to eat. However some local recipes call for the fish to be cooked and then de-boned before the flesh is incorporated into the final dish thus solving the problem.

### Preserved Fish – Cheseef

A few stalls in the market sell a variety of preserved fish, called cheseef in the UAE and recalling the state of fish consumption prior to modern communications. Mal-lah is another term for dried fish.

Two natural methods of food preservation have been available in the Emirates since prehistoric times.

Sun drying is still used to preserve prawns, anchovies, sardines and shark. The process is simple, effective and preserves excess supplies.

Salting is used for a number of fish varieties and these are still available in the suq. On the Gulf shore there are extensive salt flats where sea salt was naturally produced and collected on the sabkha salt flat between Dubai and Sharjah. There were also inland sources of salt that were collected by the bedouin and taken to the regional markets.

Dried shark, awal, is the second most common dried fish and also unmistakable as the whole shark is cut longitudinally for the drying process like the structure of a giant net. Apart from sale as whole pieces, a popular choice for travellers visiting the market, current practice is to cut the dried shark into pieces about 4 inches long, rejecting the less meaty portions and presenting the pieces in a plastic bag at a premium price. It is readily recognisable due to its characteristic pungent smell of ammonia. The ammonia smell is due to the presence of urea in the shark flesh. Sharks and rays doe not have kidneys and are hence unable to dispose of urea which builds up in their flesh.

The tuna is also cut into smaller pieces for sale and there are whole small seabreams. The dried shrimps are quite small specimens and are intact. Dried shellfish are also on sale, khart, and were described as being dried oysters. Shellfish are not common in the market despite their abundance in the marine environment. There is a religious proscription against shellfish, however given the number of oysters that had to be processed in the pearling industry, it would seem logical for some use to have been made of this abundant food.

### *Fish Sauces – Meshawaa*

Meshawaa (mehiawah in Qatar) is a product which consumed on bread, khamir or chebab, at a demonstration of traditional breads. It certainly had the correct salty taste. It was introduced from Iran to the Gulf countries. It has not entered commercial production. Preparation is from pickled Indian oil sardines, oom, water, spices and salt. These are mixed and left to ferment in a glass bottle in the sun for one to two weeks. The contents are then mashed and mixed with roasted spices to undergo a further fermentation. The sauce is spread on flatbreads and particularly eaten at breakfast with spring onions.

Tareeh is home prepared from dried oom, again fermented with salt, cummin and red chillies. It is more concentrated than meshawaa and is diluted with water for consumption on bread with radish tops and spring onions.

### *Local Recipes*

There are few accounts of local food published in English. Reviewing recipes that are available for traditional food in the UAE and neighbouring countries reveals several categories of fish dishes.

Traditional food is the food endemic in the UAE before the advent of oil wealth and the social changes which ensued. This was already a cultural mix with Iranian, sub-continental, Iraqi and north Arabian influences. It had however developed a distinctive character which is evident in the fish dishes, particularly those requiring preserved products.

Given the good quality of fish, the simple methods of frying and grilling fish, samak maqli and meshwi, are understandably popular. The ready availability of cooking oils is relatively recent and the frying medium was probably clarified butter, samn. Ovens were the clay tanoor type and do not seem to have been used for cooking fish dishes. Fish stews are well represented with spices being an essential component. The local spice mix, bezar, was used for meat and fish dishes.

Dried anchovies or sardines were ground with roasted fennel seeds to make a garnish called sahnah for rice. Gashr are also cooked with egg and cheese and eaten with bread. In Oman red pepper and garlic were pounded with the anchovies to make a similar condiment. Awal was used for a range of dishes including stews and salads. It was prepared by soaking and boiling before being incorporated into the dish. The recipes for this product are almost all Omani and this reinforces my view that these products are falling out of favour in the UAE. Local advice is to avoid drinking milk after eating awal and other dried fish as it is likely to upset the stomach. Oman also has many recipes for salted fish, malih, and a keen general appreciation of its broadly based and historic culture.

Matharubah originates from Kuwait and is made from fish and rice which are reduced to a paste after the first stage of cooking. There are recipes for fish kebabs and fishcakes and many recipes include a stage where the fish skin and bones are removed and the fish is flaked into the dish. This could explain the fish butchery in the suq.

Some Omani recipes call for the dish to be smoked as a final cooking stage. This is achieved by placing half a lime rind on the surface of the dish, placing a little samn in it, adding a piece of burning charcoal and sealing the lid so that the charcoal smolders and flavours the food. This seems to be a particular taste of Oman.

Crustaceans were also popular food and the Gulf prawns have a well deserved culinary reputation. Swimming crabs are a common market item, the remaining coastal mangroves providing the necessary breeding environment. A national told me that they were one of his childrens' favourite foods, simply cooked on the barbecue. Dugong meat was also eaten when these mammals were caught in the fishing nets and was considered a delicacy. These mammals may weigh up to 500 Kg. and the flesh is like very tender beef.

### From Fresh Water to Salt Water

Freshwater fish are fish that spend some or all of their lives in fresh water, such as rivers and lakes, with a salinity of less than 0.05%. These environments differ from marine conditions in many ways, the most obvious being the difference in levels of salinity. To survive fresh water, the fish need a range of physiological adaptations.

41.24% of all known species of fish are found in fresh water. This is primarily due to the rapid speciation that the scattered habitats make possible. When dealing with ponds and lakes, one might use the same basic models of speciation as when studying island biogeography.

### Physiology

Freshwater fish differ physiologically from salt water fish in several respects. Their gills must be able to diffuse dissolved gasses while keeping the salts in the body fluids inside. Their scales reduce water diffusion through the skin: freshwater fish that have lost too many scales will die. They also have well developed kidneys to reclaim salts from body fluids before excretion.

### Migrating Fish

Many species of fish do reproduce in freshwater, but spend most of their adult lives in the sea. These are known as anadromous fish, and include, for instance, salmon, trout and three-spined stickleback. Some other kinds of fish are, on the contrary, born in salt water, but live most of or parts of their adult lives in fresh water; for instance the eels. These are known as catadromous fish.

Species migrating between marine and fresh waters need adaptations for both environments; when in salt water they need to keep the bodily salt concentration on a level lower than the surroundings, and vice versa.

Many species solve this problem by associating different habitats with different stages of life. Both eels, anadromous salmoniform fish and the sea lamprey have different tolerances in salinity in different stages of their lives.

### Status

North America:

About four in ten North American freshwater fish are endangered, according to a pan-North American study, the main cause being human pollution. The number of fish species and subspecies to become endangered has risen from 40 to 61, since 1989.

## Saltwater Fishing

Saltwater fish are fish that spend some or all of their lives in salt water such as oceans or salt lakes, generally with a salinity of more than 0.05%.

### *Categorisation of Saltwater Fish by Habitats*

- Coastal fish (also offshore fish or neritic fish) inhabit the sea between the shoreline and the edge of the continental shelf
- Deep sea fish live below the photic zone of the ocean, i.e. where not enough light penetrates for photosynthesis to occur
- Pelagic fish live near the surface of the sea or a lake
- Demersal fish live on or near the bottom of the sea or a lake
- Coral reef fish are associated with a coral reef.

Saltwater fishing is as much about the adventure as it is about the fish. Many of the sport fish species can be big and mean, and the water can be big and bad. From shallow saltwater flats to deep-ocean fishing, saltwater anglers chase everything from dainty speckled trout to massive blue marlin in some of the most intense and inspirational surroundings on earth.

| *Principal commercial tilapia species* | | | | | | | | | | |
|---|---|---|---|---|---|---|---|---|---|---|
| *Common name* | *Scientific name* | *Maximum length* | *Common length* | *Maximum weight* | *Maximum age* | *Trophic level* | *Fish Base* | *FAO* | *WoRMS* | *IUCN status* |
| Nile tilapia | *Oreochromis niloticus* (Linnaeus, 1758) | 60 cm | cm | 4.324 kg | 9 years | 2.0 | | | | Not assessed |
| [ Blue tilapia ] | - *Oreochromis aureus* (Steindachner, 1864) | 45.7 cm | 16 cm | 2.010 kg | years | 2.1 | | | | Least concern[23] |
| Nile tilapia + blue tilapia hybrid | | cm | cm | kg | years | | | | | |
| Mozambique tilapia | *Oreochromis mossambicus* (Peters, 1852) | 39 cm | 35 cm | 1.130 kg | 11 years | 2.0 | | | | Near threatened |

Global harvest of tilapia species in million tonnes as reported by the FAO, 1950–2009

Farmed tilapia production is about 1,500,000 tonnes (1,500,000 long tons; 1,700,000 short tons) annually with an estimated value of US$1.8 billion, about equal to that of salmon and trout. Unlike carnivorous fish, tilapia can feed on algae or any plant-based food. This reduces the cost of tilapia farming, reduces fishing pressure on prey species, avoids concentrating toxins that accumulate at higher levels of the food chain and makes tilapia the preferred "aquatic chickens" of the trade.

Because of their large size, rapid growth, and palatability, tilapiine cichlids are the focus of major farming efforts, specifically various species of *Oreochromis*, *Sarotherodon*, and *Tilapia*, collectively known colloquially as tilapia. Like other large fish, they are a good source of protein and popular among artisanal and commercial fisheries. Most such fisheries were originally found in Africa, but outdoor fish farms in tropical countries, such as Papua New Guinea, the Philippines, and Indonesia, are underway in freshwater lakes. In temperate zone localities, tilapiine farming operations require energy to warm the water to tropical temperatures. One method uses waste heat from factories and power stations. China is the largest tilapia producer in the world, followed by Egypt. In modern aquaculture, wild-type Nile tilapia are not too often seen, as the dark colour of their flesh is not much desired by many customers, and because it has a bit of a reputation of being a trash fish associated with poverty. On the other hand, they are fast-growing and give good fillets; leucistic ("Red") breeds which have lighter meat have been developed and these are very popular.

Hybrid stock is also used in aquaculture; Nile × blue tilapia hybrids are usually rather dark, but a light-coloured hybrid breed known as "Rocky Mountain White" tilapia is often grown due to its very light flesh and tolerance of low temperatures. Commercially grown tilapia are almost exclusively male. Cultivators use hormones, such as testosterone, to reverse the sex of newly spawned females. Because tilapia are prolific breeders, the presence of female tilapia results in rapidly increasing populations of small fish, rather than a stable population of harvest-size animals.

Other methods of tilapia population control are polyculture, with predators farmed alongside tilapia or hybridization with other species.

### As Food

Whole tilapia fish can be processed into skinless, boneless (Pin-Bone Out, or PBO) fillets: the yield is from 30 percent to 37 percent, depending on fillet size and final trim. The use of tilapia in the commercial food industry has led to the virtual extinction of genetically pure bloodlines. Most wild tilapia today are hybrids of several species. Tilapia have very low levels of mercury, as they are fast-growing, lean and short-lived, with a primarily vegetarian diet, so do not accumulate mercury found in prey. Feral tilapia, however, may accumulate substantial quantities of mercury. Tilapia are low in saturated fat, calories, carbohydrates and sodium, and are a good protein source. They also contain the micronutrients phosphorus, niacin, selenium, vitamin $B_{12}$ and potassium.

However, typical farm-raised tilapia (the least expensive and most popular source) have low levels of omega-3 fatty acids (the essential nutrient that is an important reason that dieticians recommend eating fish), and a relatively high proportion of omega-6. "Ratios of long-chain omega-6 to long-chain omega-3, AA to EPA, respectively, in tilapia averaged about 11:1, compared to much less than 1:1 (indicating more EPA than AA) in both salmon and trout," reported a study published in July 2008. The report suggests the nutritional value of farm-raised tilapia may be compromised by the amount of corn included in the feed. The corn contains short-chain omega-6 fatty acids that contribute to the buildup of these materials in the fish.

The lower amounts of omega-3 and the higher ratios of omega-6 fats in US-farmed tilapia raised questions about the health benefits of consuming farmed tilapia fish. Some media reports even controversially suggested that farm-raised tilapia may be worse for the heart than eating bacon or a hamburger. This prompted the release of an open letter, signed by 16 science and health experts from around the world, that stated that both oily (i.e. high in omega-3 fatty acids) fish and lean fish like tilapia are an important part of the diet and concluded that "replacing tilapia or catfish with 'bacon, hamburgers or doughnuts' is absolutely not recommended."

Multiple studies have evaluated the effects of adding flaxseed derivatives (a vegetable source of omega-3 fatty acids) to the feed of farmed tilapia. These studies have found both the more common omega-3 fatty acid found in the flax, ALA and the two types almost unique to animal sources (DHA and EPA), increased in the fish fed this diet. Guided by these findings, tilapia farming techniques could be adjusted to address the nutritional criticisms directed at the fish while retaining its advantage as an omnivore capable of feeding on economically and environmentally inexpensive vegetable protein. Adequate diets for salmon and other carnivorous fish can alternatively be formulated from protein sources such as soybean, although soy-based diets may also change in the balance between omega-6 and omega-3 fatty acids.

The US produced 1.5 million tons of tilapia in 2005, with 2.5 million projected by 2010.

### *Miscellaneous Uses*

Tilapia serve as a natural, biological control for most aquatic plant problems. Tilapia consume floating aquatic plants, such as duckweed watermeal (*Lemna* sp.), most "undesirable" submerged plants, and most forms of algae. In the United States and countries such as Thailand, they are becoming the plant control method of choice, reducing or eliminating the use of toxic chemicals and heavy metal-based algaecides. Tilapia rarely compete with other "pond" fish for food. Instead, because they consume plants and nutrients unused by other fish species and substantially reduce oxygen-depleting detritus; adding tilapia often increases the population, size and health of other fish. Arizona stocks tilapia in the canals that serve as the drinking water sources for the cities of Phoenix, Mesa and others. The fish help purify the water by consuming vegetation and detritus, greatly reducing purification costs. Arkansas stocks many public ponds and lakes to help with vegetation control, favouring tilapia as a robust forage species and for anglers.

In Kenya, tilapia help control mosquitoes which carry malaria parasites. They consume mosquito larvae, which reduces the numbers of adult females, the disease's vector. Tilapia also provide an abundant food source for aquatic predators.

## Krill Oil

Krill oil is made from a species of krill [*Euphausia superba*]. Two of the most important nutrients in krill oil are:

(1) omega-3 fatty acids similar to those of fish oil and

(2) Phospholipid-derived fatty acids (PLFA), mainly phosphatidylcholine (alternatively referred to as marine lecithin).

Also, antioxidant experimental egg products with krill oil likely contained astaxanthin, a natural antioxidant. The fatty acid composition in the phospholipids in krill oil has been described in two papers.

Several studies have shown toxic residues in Antarctic krill and fish.

### *Krill Oil vs. Fish Oil*

Considering the much higher price for krill oil, the potentially small increase in bioavailability may not be worth it. Until data exists comparing fish oil to krill oil on intermediate markers of risk and actual disease endpoints it will be difficult to say one is better than the other.

### *Ecological Concerns*

The harvesting of Antarctic krill is relatively new. The vast majority is harvested for feed for fish farms. A small percentage (2 percent in the 2009-2010 season) is harvested for human consumption.

Krill is considered by many scientists to be the largest biomass in the world. Antarctic krill is fundamental to the survival of almost every species of animal that lives in the Antarctic or sub-Antarctic waters and island groups. Because Antarctic krill are so important, in 1982, the United States, the United Kingdom, Australia, South Africa, New Zealand, Chile, European Community, Germany and Japan formed a treaty organisation to ensure that krill were being harvested sustainably. Named the Commission for the Conservation of Antarctic Marine Living Resources (CCAMLR-pronounced "camel-lahr"), it now manages the fin fish (mostly toothfish) and krill fisheries in the Southern Ocean. Scientists from some of the CCAMLR member nations, including Australia, the United States and the United Kingdom, conduct research in the Southern Ocean and make recommendations to CCAMLR that enable the organisation to make management decisions. Currently there are 25 Members of CCAMLR, 24 member states and the European Community.

CCAMLR has successfully implemented a precautionary and ecosystem approach in managing the krill fishery. The krill fishery is considered by some of those scientists to be among the best managed fisheries in the world, providing strict catch limits on licensed vessels, and scientific observers on board. CCAMLR scientists are working to take into account the possible effects of climate change on the ecosystem as well as effects of changes in technology and operational pattern of krill fishing vessels on the fishery when making management decisions.

# 7

# Seafood on American Menus: Past and Present

Seafood is any form of sea life regarded as food by humans. Seafood prominently includes fish and shellfish. Shellfish include various species of molluscs, crustaceans, and echinoderms. Historically, sea mammals such as whales and dolphins have been consumed as food, though that happens to a lesser extent these days. Edible sea plants, such as some seaweeds and microalgae, are widely eaten as seafood around the world, especially in Asia. In North America, although not generally in the United Kingdom, the term "seafood" is extended to fresh water organisms eaten by humans, so all edible aquatic life may be referred to as seafood. For the sake of completeness, this article includes all edible aquatic life.

The harvesting of wild seafood is known as fishing and the cultivation and farming of seafood is known as aquaculture, mariculture, or in the case of fish, fish farming. Seafood is often distinguished from meat, although it is still animal and is excluded in a strict vegetarian diet. Seafood is an important source of protein in many diets around the world, especially in coastal areas.

Most of the seafood harvest is consumed by humans, but a significant proportion is used as fish food to farm other fish or rear farm animal. Some seafoods (kelp) are used as food for other plants (fertilizer). In these ways, seafoods are indirectly used to produce further food for human consumption. Products, such as fish oil and spirulina tablets are also extracted from seafoods. Some seafood is feed to aquarium fish, or used to feed domestic pets, such as cats, and a small proportion is used in medicine, or is used industrially for non-food purposes (leather).

***Figure:*** *Seafood includes any form of food taken from the sea*

## The Salmon of the River Wye

Salmon /ÈsæmYn/ is the common name for several species of fish in the family Salmonidae. Other fish in the same family are called trout. The difference between salmon and trout is sometimes said to be that salmon migrate and trout are resident, and that salmon spawn once and trout spawn many times, though these distinctions are not always strictly true. Salmon live along the coasts of both the North Atlantic (the migratory species *Salmo salar*) and Pacific Oceans (half a dozen species of the genus *Oncorhynchus*), and have also been introduced into the Great Lakes of North America. Salmon are intensively produced in aquaculture in many parts of the world.

Typically, salmon are anadromous: they are born in fresh water, migrate to the ocean, then return to fresh water to reproduce. However, populations of several species are restricted to fresh water through their lives. Folklore has it that the fish return to the exact spot where they were born to spawn; tracking studies have shown this to be true, and this homing behaviour has been shown to depend on olfactory memory.

## *Species*

The term "salmon" comes from the Latin *salmo*, which in turn may have originated from *salire*, meaning "to leap". The nine commercially important species of salmon occur in two genera. The genus *Salmo* contains the Atlantic salmon, found in the north Atlantic. The genus *Oncorhynchus* contains eight species which occur naturally only in the north Pacific. Chinook salmon have been introduced in New Zealand. As a group, these are known as Pacific salmon.

| *Atlantic and Pacific salmon* | | | | | | | | | | | |
|---|---|---|---|---|---|---|---|---|---|---|---|
| ***Genus*** | ***Common name*** | ***Scientific name*** | ***Maximum length*** | ***Common length*** | ***Maximum weight*** | ***Maximum age*** | ***Trophic level*** | ***Fish Base*** | ***FAO*** | ***ITIS*** | ***IUCN status*** |
| *Salmo* (Atlantic salmon) | Atlantic salmon | *Salmo salar* Linnaeus, 1758 | 150 cm | 120 cm | 46.8 kg | 13 years | 4.4 | | | | Least concern |
| *Oncorhynchus* (Pacific salmon) | Chinook salmon | *Oncorhynchus tshawytscha* (Walbaum, 1792) | 150 cm | 70 cm | 61.4 kg | 9 years | 4.4 | | | | Not assessed |
| | Chum salmon | *Oncorhynchus keta* (Walbaum, 1792) | 100 cm | 58 cm | 15.9 kg | 7 years | 3.5 | | | | Not assessed |
| | Coho salmon | *Oncorhynchus kisutch* (Walbaum, 1792) | 108 cm | 71 cm | 15.2 kg | 5 years | 4.2 | | | | Not assessed |
| | Pink salmon | *Oncorhynchus gorbuscha* (Walbaum, 1792) | 76 cm | 50 cm | 6.8 kg | 3 years | 4.2 | | | | Not assessed |
| | Sockeye salmon | *Oncorhynchus nerka* (Walbaum, 1792) | 84 cm | 58 cm | 7.7 kg | 8 years | 3.7 | | | | Least concern[24] |
| | Steelhead† (rainbow trout) | *Oncorhynchus mykiss* (Walbaum, 1792) | 79.0 cm | cm | 10.0 kg | years | 3.6 | | | | Not assessed |
| | Masu salmon | *Oncorhynchus masou* (Brevoort, 1856) | 79.0 cm | cm | 10.0 kg | 3 years | 3.6 | | | | Not assessed |

Both the *Salmo* and *Oncorhynchus* genera also contain a number of species referred to as trout. Within *Salmo*, additional minor taxa have been called salmon in English, i.e. the Adriatic salmon (*Salmo obtusirostris*) and Black Sea salmon (*Salmo labrax*).

The steelhead morph of the rainbow trout migrates to sea, but it is not termed "salmon".

There are also a number of other species whose common names refer to them as being salmon. Of those listed below, the Danube salmon or *huchen* is a large freshwater salmonid related to the salmon above, but others are marine fishes of the non-related perciform-order:

| *Some other fishes called salmon* | | | | | | | | | | |
|---|---|---|---|---|---|---|---|---|---|---|
| *Comm on name* | *Scientific name* | *Maxim um length* | *Comm on length* | *Maxim um weight* | *Maxim um age* | *Trop hic level* | *Fis h Ba se* | *FA O* | *ITI S* | *IUCN status* |
| Danube salmon | *Hucho hucho* (Linnaeus, 1758) | 150 cm | 70 cm | 52 kg | 15 years | 4.2 | | | | Endange red |
| Austral ian salmon | *Arripis trutta* (Forster, 1801) | 89 cm | 47 cm | 9.4 kg | 26 years | 4.1 | | | | Not assessed |
| Hawaii an salmon | *Elagatis bipinnulata* (Quoy & Gaimard, 1825) | 180 cm | 90 cm | 46.2 kg | years | 3.6 | | | | Not assessed |
| Indian salmon | *Eleutheronema tetradactylum* (Shaw, 1804) | 200 cm | 50 cm | 145 kg | years | 4.4 | | | | Not assessed |

*Eosalmo driftwoodensis*, the oldest known salmon in the fossil record, helps scientists figure how the different species of salmon diverged from a common ancestor. The British Columbia salmon fossil provides evidence that the divergence between Pacific and Atlantic salmon had not yet occurred 40 million years ago. Both the fossil record and analysis of mitochondrial DNA suggest the divergence occurred by 10 to 20 million years ago. This independent evidence from DNA analysis and the fossil record reject the glacial theory of salmon divergence.

## Distribution

***Figure:*** *Atlantic salmon,* Salmo salar

- Atlantic salmon (*Salmo salar*) reproduces in northern rivers on both coasts of the Atlantic Ocean.
  - o Landlocked salmon (*Salmo salar* m. *sebago*) live in a number of lakes in eastern North America and in Northern Europe, for instance in lakes Onega, Ladoga, Saimaa, Vänern and Winnipesaukee. They are not a different species from the Atlantic salmon, but have independently evolved a non-migratory life cycle, which they maintain even when they could access the ocean.
- Masu salmon or cherry salmon (*Oncorhynchus masou*) is found only in the western Pacific Ocean in Japan, Korea and Russia. A land-locked subspecies known as the Taiwanese salmon or Formosan salmon (*Oncorhynchus masou formosanus*) is found in central Taiwan's Chi Chia Wan Stream.
- Chinook salmon (*Oncorhynchus tshawytscha*) is also known in the US as king salmon or blackmouth salmon, and as spring salmon in British Columbia. Chinook are the largest of all Pacific salmon, frequently exceeding 30 lb (14 kg). The name Tyee is used in British Columbia to refer to Chinook over 30 pounds, and in Columbia River watershed, especially large Chinook were once referred to as June hogs. Chinook salmon are known to range as far north as the Mackenzie River and Kugluktuk in the central Canadian arctic, and as far south as the Central California Coast.
- Chum salmon (*Oncorhynchus keta*) is known as dog, keta, or calico salmon in some parts of the US. This species has the widest geographic range of the Pacific species: south to the Sacramento River in California in the eastern Pacific and the island of Kyûshû in the Sea of Japan in the western Pacific; north to the Mackenzie River in Canada in the east and to the Lena River in Siberia in the west.
- Coho salmon (*Oncorhynchus kisutch*) is also known in the US as silver salmon. This species is found throughout the coastal waters of Alaska and British Columbia and as far south as Central California (Monterey Bay). It is also now known to occur, albeit infrequently, in the Mackenzie River.
- Pink salmon (*Oncorhynchus gorbuscha*), known as humpies in southeast and southwest Alaska, are found from northern California and Korea, throughout the northern Pacific, and from the Mackenzie River in Canada to the Lena River in

Siberia, usually in shorter coastal streams. It is the smallest of the Pacific species, with an average weight of 3.5 to 4.0 lb (1.6 to 1.8 kg).

- Sockeye salmon (*Oncorhynchus nerka*) is also known in the US as red salmon. This lake-rearing species is found south as far as the Klamath River in California in the eastern Pacific and northern Hokkaidô island in Japan in the western Pacific and as far north as Bathurst Inlet in the Canadian Arctic in the east and the Anadyr River in Siberia in the west. Although most adult Pacific salmon feed on small fish, shrimp and squid; sockeye feed on plankton they filter through gill rakers. Kokanee salmon is a land-locked form of sockeye salmon.
- The Danube salmon or huchen (*Hucho hucho*), is the largest permanent fresh water salmonid species.

### Life Cycle

Salmon eggs are laid in freshwater streams typically at high latitudes. The eggs hatch into alevin or sac fry. The fry quickly develop into parr with camouflaging vertical stripes. The parr stay for six months to three years in their natal stream before becoming smolts, which are distinguished by their bright, silvery colour with scales that are easily rubbed off. Only 10% of all salmon eggs are estimated to survive to this stage.

The smolt body chemistry changes, allowing them to live in saltwater. Smolts spend a portion of their out-migration time in brackish water, where their body chemistry becomes accustomed to osmoregulation in the ocean.

***Figure:*** *Juvenile salmon, parr, grow up in the relatively protected natal river*

The salmon spend about one to five years (depending on the species) in the open ocean, where they gradually become sexually mature. The adult salmon then return primarily to their natal streams to spawn. In Alaska, crossing over to other streams allows salmon to populate new streams, such as those that emerge as a glacier retreats. The precise method salmon use to navigate has not been established, though their keen sense of smell is involved. Atlantic salmon spend between one and four years at sea. (When a fish returns after just one year's sea feeding, it is called a grilse in Canada, Britain and Ireland.) Prior to spawning, depending on the species, salmon undergo changes. They may grow a hump, develop canine teeth, develop a kype (a pronounced curvature of the jaws in male salmon).

All will change from the silvery blue of a fresh-run fish from the sea to a darker colour. Salmon can make amazing journeys, sometimes moving hundreds of miles upstream against strong currents and rapids to reproduce. Chinook and sockeye salmon from central Idaho, for example, travel over 900 miles (1,400 km) and climb nearly 7,000 feet (2,100 m) from the Pacific Ocean as they return to spawn. Condition tends to deteriorate the longer the fish remain in fresh water, and they then deteriorate further after they spawn, when they are known as kelts. In all species of Pacific salmon, the mature individuals die within a few days or weeks of spawning, a trait known as semelparity. Between 2 and 4% of Atlantic salmon kelts survive to spawn again, all females. However, even in those species of salmon that may survive to spawn more than once (iteroparity), postspawning mortality is quite high (perhaps as high as 40 to 50%.)

To lay her roe, the female salmon uses her tail (caudal fin), to create a low-pressure zone, lifting gravel to be swept downstream, excavating a shallow depression, called a redd. The redd may sometimes contain 5,000 eggs covering 30 square feet (2.8 $m^2$). The eggs usually range from orange to red. One or more males will approach the female in her redd, depositing his sperm, or milt, over the roe. The female then covers the eggs by disturbing the gravel at the upstream edge of the depression before moving on to make another redd. The female will make as many as seven redds before her supply of eggs is exhausted.

Each year, the fish experiences a period of rapid growth, often in summer, and one of slower growth, normally in winter. This results in ring formation around an earbone called the otolith, (annuli) analogous to the growth rings visible in a tree trunk. Freshwater growth shows as densely crowded rings, sea growth as widely spaced rings; spawning is marked by significant erosion as body mass is converted into eggs and milt.

Freshwater streams and estuaries provide important habitat for many salmon species. They feed on terrestrial and aquatic insects, amphipods, and other crustaceans while young, and primarily on other fish when older. Eggs are laid in deeper water with larger gravel, and need cool water and good water flow (to supply oxygen) to the developing embryos. Mortality of salmon in the early life stages is usually high due to natural predation and human-induced changes in habitat, such as siltation, high water temperatures, low oxygen concentration, loss of stream cover, and reductions in river flow. Estuaries and their associated wetlands provide vital nursery areas for the salmon prior to their departure to the open ocean. Wetlands not only help buffer the estuary from silt and pollutants, but also provide important feeding and hiding areas.

Salmon not killed by other means show greatly accelerated deterioration (phenoptosis, or "programmed aging") at the end of their lives. Their bodies rapidly deteriorate right after they spawn as a result of the release of massive amounts of corticosteroids.

## Ecology

### *Bears and Salmon*

In the Pacific Northwest and Alaska, salmon are keystone species, supporting wildlife such as birds, bears and otters. The bodies of salmon represent a transfer of nutrients from the ocean, rich in nitrogen, sulfur, carbon and phosphorus, to the forest ecosystem.

Grizzly bears function as ecosystem engineers, capturing salmon and carrying them into adjacent wooded areas. There they deposit nutrient-rich urine and faeces and partially eaten carcasses. Bears are estimated to leave up to half the salmon they harvest on the forest floor, in densities that can reach 4,000 kilograms per hectare, providing as much as 24% of the total nitrogen available to the riparian woodlands. The foliage of spruce trees up to 500 m (1,600 ft) from a stream where grizzlies fish salmon have been found to contain nitrogen originating from fished salmon.

### *Beavers and Salmon*

Beavers also function as ecosystem engineers; in the process of clear-cutting and damming, beavers alter their ecosystems extensively. Beaver ponds can provide critical habitat for juvenile salmon. An example of this was seen in the years following 1818 in the Columbia River Basin. In 1818, the British government made an agreement with the U.S. government to allow U.S. citizens access to the Columbia catchment. At the time, the Hudson's Bay Company sent word to trappers to extirpate all furbearers from the area in an effort to make the area less attractive to U.S. fur traders. In response to the elimination of beavers from large parts of the river system, salmon runs plummeted, even in the absence of many of the factors usually associated with the demise of salmon runs. Salmon recruitment can be affected by beavers' dams because dams can:

- Slow the rate at which nutrients are flushed from the system; nutrients provided by adult salmon dying throughout the fall and winter remain available in the spring to newly-hatched juveniles
- Provide deeper water pools where young salmon can avoid avian predators
- Increase productivity through photosynthesis and by enhancing the conversion efficiency of the cellulose-powered detritus cycle
- Create low-energy environments where juvenile salmon put the food they ingest into growth rather than into fighting currents
- Increase structural complexity with many physical niches where salmon can avoid predators

Beavers' dams are able to nurture salmon juveniles in estuarine tidal marshes where the salinity is less than 10 ppm. Beavers build small dams of generally less than 2 feet (60 cm) high in channels in the myrtle zone. These dams can be overtopped at high tide and hold water at low tide. This provides refuges for juvenile salmon so they do not have to swim into large channels where they are subject to predation.

### Parasites

According to Canadian biologist Dorothy Kieser, the myxozoan parasite *Henneguya salminicola* is commonly found in the flesh of salmonids. It has been recorded in the field samples of salmon returning to the Haida Gwaii Islands. The fish responds by walling off the parasitic infection into a number of cysts that contain milky fluid. This fluid is an accumulation of a large number of parasites.

*Henneguya* and other parasites in the myxosporean group have complex life cycles, where the salmon is one of two hosts. The fish releases the spores after spawning. In the *Henneguya* case, the spores enter a second host, most likely an invertebrate, in the spawning stream. When juvenile salmon migrate to the Pacific Ocean, the second host releases a stage infective to salmon. The parasite is then carried in the salmon until the next spawning cycle. The myxosporean parasite that causes whirling disease in trout has a similar life cycle. However, as opposed to whirling disease, the *Henneguya* infestation does not appear to cause disease in the host salmon — even heavily infected fish tend to return to spawn successfully.

According to Dr. Kieser, a lot of work on *Henneguya salminicola* was done by scientists at the Pacific Biological Station in Nanaimo in the mid-1980s, in particular, an overview report which states, "the fish that have the longest fresh water residence time as juveniles have the most noticeable infections. Hence in order of prevalence coho are most infected followed by sockeye, chinook, chum and pink." As well, the report says, at the time the studies were conducted, stocks from the middle and upper reaches of large river systems in British Columbia such as Fraser, Skeena, Nass and from mainland coastal streams in the southern half of B.C., "are more likely to have a low prevalence of infection." The report also states, "It should be stressed that *Henneguya*, economically deleterious though it is, is harmless from the view of public health. It is strictly a fish parasite that cannot live in or affect warm blooded animals, including man".

According to Klaus Schallie, Molluscan Shellfish Program Specialist with the Canadian Food Inspection Agency, "*Henneguya salminicola* is found in southern B.C. also and in all species of salmon. I have previously examined smoked chum salmon sides that were riddled with cysts and some sockeye runs in Barkley Sound (southern B.C., west coast of Vancouver Island) are noted for their high incidence of infestation."

Sea lice, particularly *Lepeophtheirus salmonis* and various *Caligus* species, including *C. clemensi* and *C. rogercresseyi*, can cause deadly infestations of both farm-grown and wild salmon. Sea lice are ectoparasites which feed on mucus, blood, and skin, and migrate and latch onto the skin of wild salmon during free-swimming, planktonic nauplii and copepodid larval stages, which can persist for several days. Large numbers of highly populated, open-net salmon farms can create exceptionally large concentrations of sea lice; when exposed in river estuaries containing large numbers of open-net farms, many young wild salmon are infected, and do not survive as a result. Adult salmon may survive otherwise critical numbers of sea lice, but small, thin-skinned juvenile salmon migrating to sea are highly vulnerable. On the Pacific coast of Canada, the louse-induced mortality of pink salmon in some regions is commonly over 80%.

### Wild Fisheries

#### Commercial

As can be seen from the production chart at the left, the global capture reported by different countries to the FAO of commercial wild salmon has remained fairly steady since 1990 at about one million tonnes per year. This is in contrast to farmed salmon (below) which has increased in the same period from about 0.6 million tonnes to well over two million tonnes.

Nearly all captured wild salmon are Pacific salmon. The capture of wild Atlantic salmon has always been relatively small, and has declined steadily since 1990. In 2011 only 2,500 tonnes were reported. In contrast about half of all farmed salmon are Atlantic salmon.

#### Recreational

Recreational salmon fishing can be a technically demanding kind of sport fishing, not necessarily congenial for beginning fishermen. A conflict exists between commercial fishermen and recreational fishermen for the right to salmon stock resources. Commercial fishing in estuaries and coastal areas is often restricted so enough salmon can return to their natal rivers where they can spawn and be available for sport fishing. On parts of the North American west coast sport salmon fishing completely replaces inshore commercial fishing. The commercial value of a salmon can be several times less than the value of the same fish caught by a sport fisherman. This is "a powerful economic argument for allocating stock resources preferentially to sport fishing."

### Farmed Salmon

Salmon aquaculture is a major contributor to the world production of farmed finfish, representing about US$10 billion annually. Other commonly cultured fish species include: tilapia, catfish, sea bass, carp and bream. Salmon farming is significant in Chile, Norway, Scotland, Canada and the Faroe Islands, and is the source for most salmon consumed in America and Europe. Atlantic salmon are also, in very small volumes, farmed in Russia and the island of Tasmania, Australia.

Salmon are carnivorous and are currently fed a meal produced from catching other wild fish and other marine organisms. Salmon farming leads to a high demand for wild forage fish. Salmon require large nutritional intakes of protein, and consequently, farmed salmon consume more fish than they generate as a final product. To produce one pound of farmed salmon, products from several pounds of wild fish are fed to them. As the salmon farming industry expands, it requires more wild forage fish for feed, at a time when 75% of the world's monitored fisheries are already near to or have exceeded their maximum sustainable yield. The industrial-scale extraction of wild forage fish for salmon farming then impacts the survivability of the wild predator fish which rely on them for food.

Work continues on substituting vegetable proteins for animal proteins in the salmon diet. Unfortunately, though, this substitution results in lower levels of the highly valued omega-3 fatty acid content in the farmed product.

Intensive salmon farming now uses open-net cages, which have low production costs, but have the drawback of allowing disease and sea lice to spread to local wild salmon stocks.

On a dry weight basis, 2–4 kg of wild-caught fish are needed to produce one kg of salmon.

Another form of salmon production, which is safer but less controllable, is to raise salmon in hatcheries until they are old enough to become independent. They are then released into rivers, often in an attempt to increase the salmon population. This system is referred to as ranching, and was very common in countries such as Sweden before the Norwegians developed salmon farming, but is seldom done by private companies, as anyone may catch the salmon when they return to spawn, limiting a company's chances of benefiting financially from their investment.

Because of this, the method has mainly been used by various public authorities and nonprofit groups such as the Cook Inlet Aquaculture Association as a way of artificially increasing salmon populations in situations where they have declined due to overharvesting, construction of dams, and habitat destruction or fragmentation. Unfortunately, there can be negative consequences to this sort of population manipulation, including genetic "dilution" of the wild stocks, and many jurisdictions are now beginning to discourage supplemental fish planting in favour of harvest controls and habitat improvement and protection. A variant method of fish stocking, called ocean ranching, is under development in Alaska. There, the young salmon are released into the ocean far from any wild salmon streams.

When it is time for them to spawn, they return to where they were released where fishermen can then catch them.

An alternative method to hatcheries is to use spawning channels. These are artificial streams, usually parallel to an existing stream with concrete or rip-rap sides and gravel bottoms. Water from the adjacent stream is piped into the top of the channel, sometimes via a header pond, to settle out sediment. Spawning success is often much better in channels than in adjacent streams due to the control of floods, which in some years can wash out the natural redds. Because of the lack of floods, spawning channels must sometimes be cleaned out to remove accumulated sediment. The same floods which destroy natural redds also clean them out. Spawning channels preserve the natural selection of natural streams, as there is no benefit, as in hatcheries, to use prophylactic chemicals to control diseases.

Farm-raised salmon are fed the carotenoids astaxanthin and canthaxanthin to match their flesh colour to wild salmon. One proposed alternative to the use of wild-caught fish as feed for the salmon, is the use of soy-based products. This should be better for the local environment of the fish farm, but producing soy beans has a high environmental cost for the producing region.

Another possible alternative is a yeast-based coproduct of bioethanol production, proteinaceous fermentation biomass. Substituting such products for engineered feed can result in equal (sometimes enhanced) growth in fish. With its increasing availability, this would address the problems of rising costs for buying hatchery fish feed. Yet another attractive alternative is the increased use of seaweed. Seaweed provides essential minerals and vitamins for growing organisms. It offers the advantage of providing natural amounts of dietary fiber and having a lower glycemic load than grain-based fish meal. In the best-case scenario, widespread use of seaweed could yield a future in aquaculture that eliminates the need for land, freshwater, or fertilizer to raise fish.

### *Management*

The population of wild salmon declined markedly in recent decades, especially North Atlantic populations, which spawn in the waters of western Europe and eastern Canada, and wild salmon in the Snake and Columbia River systems in northwestern United States.

Salmon population levels are of concern in the Atlantic and in some parts of the Pacific. Alaska fishery stocks are still abundant, and catches have been on the rise in recent decades, after the state initiated limitations in 1972. Some of the most important Alaskan salmon sustainable wild fisheries are located near the Kenai River, Copper River, and in Bristol Bay. Fish farming of Pacific salmon is outlawed in the United States Exclusive Economic Zone, however, there is a substantial network of publicly funded hatcheries, and the State of Alaska's fisheries management system is viewed as a leader in the management of wild fish stocks.

In Canada, returning Skeena River wild salmon support commercial, subsistence and recreational fisheries, as well as the area's diverse wildlife on the coast and around communities hundreds of miles inland in the watershed. The status of wild salmon in Washington is mixed. Of 435 wild stocks of salmon and steelhead, only 187 of them were classified as healthy; 113 had an unknown status, one was extinct, 12 were in critical condition and 122 were experiencing depressed populations.

The commercial salmon fisheries in California have been either severely curtailed or closed completely in recent years, due to critically low returns on the Klamath and or Sacramento Rivers, causing millions of dollars in losses to commercial fishermen. Both Atlantic and Pacific salmon are popular sportfish.

Salmon populations now exist in all the Great Lakes. Coho stocks were planted in the late 1960s in response to the growing population of non-native alewife by the state of Michigan. Now Chinook (king), Atlantic, and coho (silver) salmon are annually stocked in all Great Lakes by most bordering states and provinces. These populations are not self-sustaining and do not provide much in the way of a commercial fishery, but have led to the development of a thriving sport fishery.

### As Food

Salmon is a popular food. Classified as an oily fish, salmon is considered to be healthy due to the fish's high protein, high omega-3 fatty acids, and high vitamin D content. Salmon is also a source of cholesterol, with a range of 23–214 mg/100 g depending on the species. According to reports in the journal *Science*, however, farmed salmon may contain high levels of dioxins. PCB (polychlorinated biphenyl) levels may be up to eight times higher in farmed salmon than in wild salmon, but still well below levels considered dangerous. Nonetheless, according to a 2006 study published in the Journal of the American Medical Association, the benefits of eating even farmed salmon still outweigh any risks imposed by contaminants. The type of omega-3 present may not be a factor for other important health functions.

The vast majority of Atlantic salmon available around the world are farmed (almost 99%), whereas the majority of Pacific salmon are wild-caught (greater than 80%). Canned salmon in the US is usually wild Pacific catch, though some farmed salmon is available in canned form. Smoked salmon is another popular preparation method, and can either be hot or cold smoked. Lox can refer to either cold-smoked salmon or salmon cured in a brine solution (also called gravlax). Traditional canned salmon includes some skin (which is harmless) and bone (which adds calcium). Skinless and boneless canned salmon is also available. Raw salmon flesh may contain *Anisakis* nematodes, marine parasites that cause anisakiasis. Before the availability of refrigeration, the Japanese did not consume raw salmon. Salmon and salmon roe have only recently come into use in making sashimi (raw fish) and sushi.

### Colour

Salmon flesh is generally orange to red, although there are some examples of white-fleshed wild salmon. The natural colour of salmon results from carotenoid pigments, largely astaxanthin but also canthaxanthin, in the flesh. Wild salmon get these carotenoids from eating krill and other tiny shellfish. Because consumers have shown a reluctance to purchase white-fleshed salmon, astaxanthin (E161j), and very minutely canthaxanthin (E161g), are added as artificial colorants to the feed of farmed salmon, because prepared diets do not naturally contain these pigments.

In most cases, the astaxanthin is made chemically; alternatively it is extracted from shrimp flour. Another possibility is the use of dried red yeast, which provides the same pigment. However, synthetic mixtures are the least expensive option. Astaxanthin is a potent antioxidant that stimulates the development of healthy fish nervous systems and enhances the fish's fertility and growth rate. Research has revealed canthaxanthin may have negative effects on the human eye, accumulating in the retina at high levels of consumption.

Today, the concentration of carotenoids (mainly canthaxanthin and astaxanthin) exceeds 8 mg/kg of flesh, and all fish producers try to reach a level that represents a value of 16 on the "Roche Colour Card", a colour card used to show how pink the fish will appear at specific doses. This scale is specific for measuring the pink colour due to astaxanthin and is not for the orange hue obtained with canthaxanthin. The development of processing and storage operations, which can be detrimental on canthaxanthin flesh concentration, has led to an increased quantity of pigments added to the diet to compensate for the degrading effects of the processing. In wild fish, carotenoid levels of up to 25 mg are present, but levels of canthaxanthin are, in contrast, minor.

## Products

A simple rule of thumb is that the vast majority of Atlantic salmon available on the world market are farmed (almost 99%), whereas the majority of Pacific salmon are wild caught (greater than 80%). Canned salmon in the U.S. is usually wild Pacific catch, though some farmed salmon is available in canned form. Smoked salmon is another popular preparation method, and can either be hot or cold smoked. Lox can refer either to cold smoked salmon or to salmon cured in a brine solution (also called gravlax). Traditional canned salmon includes some skin (which is harmless) and bone (which adds calcium). Skinless and boneless canned salmon is also available.

Raw salmon flesh may contain *Anisakis* nematodes, marine parasites that cause Anisakiasis. Before the availability of refrigeration, Japanese did not consume raw salmon. Salmon and salmon roe have only recently come into use in making sashimi (raw fish) and sushi. Ordinary types of cooked salmon contain 500–1500 mg DHA and 300–1000 mg EPA per 100 grams

## Salmon Dishes

| *Name* | *Origin* | *Description* |
|---|---|---|
| Gravlax | Nordic | Raw salmon cured in salt, sugar, and dill. Usually served as an appetizer, sliced thinly and accompanied by *hovmästarsås* (also known as *gravlaxsås*), a dill and mustard sauce, either on bread of some kind, or with boiled potatoes. |
| Lohikeitto | Nordic | A creamy salmon soup consisting of salmon fillets, boiled potatoes and leeks, served hot with some dill. |
| Lomi salmon | Polynesian | A side dish consisting of fresh tomato and salmon salad. It was introduced to Hawaiians by early western sailors. It is typically prepared by mixing raw salted, diced salmon with tomatoes, sweet gentle Maui onions (or sometimes green onion), and occasionally flakes of hot red chili pepper, or crushed ice. It is always served cold. Other variations include salmon, diced tomato, diced cucumber, and chopped sweet onion. |

*Contd...*

| ***Name*** | ***Origin*** | ***Description*** |
|---|---|---|
| Lox | | A fillet that has been cured. In its most popular form, it is thinly sliced—less than 5 millimetres (0.20 in) in thickness—and, typically (in North America), served on a bagel, often with cream cheese, onion, tomato, cucumber and capers. Lox in small pieces is also often added and cooked into scrambled eggs, sometimes with chopped onion. |
| Poacher's Relish | | A tangy relish made with smoked salmon and lemon zest. Like Gentleman's Relish, it is usually eaten with toast, crackers or blinis. |
| Rui-be | Japan | Salmon that is frozen outdoors, sliced like sashimi, and served with soy sauce and water peppers. |
| Salmon burger | | A type of fishcake made mostly from salmon in the style of a hamburger. It is challenging to make and cook as the salmon requires a binder to make it stick together and is easy to overcook which makes it too dry. Salmon burgers are especially common in Alaska where they are routinely offered as an alternative to beef hamburgers. |
| Salmon tartare | | Appetizer prepared with fresh raw salmon and seasonings, commonly spread on a cracker or artisan style bread |
| Smoked salmon | | A preparation of salmon, typically a fillet that has been cured and then hot or cold smoked. Due to its moderately high price, smoked salmon is considered a delicacy. Although the term lox is sometimes applied to smoked salmon, they are different products. |

### History

The salmon has long been at the heart of the culture and livelihood of coastal dwellers. Many people of the northern Pacific shore had a ceremony to honour the first return of the year. For many centuries, people caught salmon as they swam upriver to spawn. A famous spearfishing site on the Columbia River at Celilo Falls was inundated after great dams were built on the river. The Ainu, of northern Japan, trained dogs to catch salmon as they returned to their breeding grounds *en masse*. Now, salmon are caught in bays and near shore.

The Columbia River salmon population is now less than 3% of what it was when Lewis and Clark arrived at the river. Salmon canneries established by settlers beginning in 1866 had a strong negative impact on the salmon population. In his 1908 State of the Union address, U.S. President Theodore Roosevelt observed that the fisheries were in significant decline:

The salmon fisheries of the Columbia River are now but a fraction of what they were twenty—five years ago, and what they would be now if the United States Government had taken complete charge of them by intervening between Oregon and Washington. During these twenty—five years the fishermen of each State have naturally tried to take all they could get, and the two legislatures have never been able to agree on joint action of any kind adequate in degree for the protection of the fisheries. At the moment the fishing on the Oregon side is practically closed, while there is no limit on the Washington side of any kind, and no one can tell what the courts will decide as to the very statutes under which this action and non—action result. Meanwhile very few salmon reach the spawning grounds, and probably four years hence the fisheries will amount to nothing; and this comes from a struggle between the associated, or gill—net, fishermen on the one hand, and the owners of the fishing wheels up the river.

## The Histamine Problem

Scombrotoxin fish poisoning (SFP), often called "histamine poisoning", is caused by ingestion of certain species of marine fish that contain high levels of histamine and possibly other biogenic amines. The fish species involved contain high levels of free histidine in their tissue and include tuna (which accounts for 8% of globally traded fish) and other pelagic species like mackerel, sardines, and anchovy, which account for significant global fish production. When these fish are subjected to temperature abuse during and/or after harvest, bacterial decarboxylation of histidine leads to histamine formation. Other biogenic amines produced during bacterial growth in fish may potentiate histamine's effect. Severity of the symptoms can vary depending on the amount of histamine and other biogenic amines ingested and the individual's sensitivity to specific biogenic amines. In some parts of the world, SFP accounts for the largest proportion of cases of fish-borne illness.

For the purposes of consumer protection, fish importing countries have regulations and varying limits for histamine in fish and fishery products. Codex Alimentarius through its standards and guidelines aims to provide countries with the basis for which to manage issues such as histamine formation. For example, the Codex Code of Practice for Fish and Fishery Products provides guidance on fish handling practices that need to be implemented to minimize food safety problems including SFP. In addition, the Codex Alimentarius has established several standards which include maximum levels for histamine in different fish and fishery products.

Different limits have been established as indicators of decomposition and as indicators of hygiene and handling. However, the associated guidance on the relevant sampling plans and other aspects of sampling have been limited or even non-existent. Furthermore, many of these limits were established in a pre-risk assessment era and the scientific basis for the limits is unclear. As food safety management moves towards more risk and evidence based approaches, there is a need to review existing limits in light of the most up to date scientific information and ensure there is a robust scientific basis for any limits recommended by Codex.

## From Lake and Sea Goldûsh and Mantis Shrimp Sushi

The goldfish (*Carassius auratus auratus*) is a freshwater fish in the family Cyprinidae of order Cypriniformes. It was one of the earliest fish to be domesticated, and is one of the most commonly kept aquarium fish.

A relatively small member of the carp family (which also includes the koi carp and the crucian carp), the goldfish is a domesticated version of a less-colourful carp (*Carassius auratus*) native to east Asia. It was first domesticated in China more than a thousand years ago, and several distinct breeds have since been developed. Goldfish breeds vary greatly in size, body shape, fin configuration and colouration (various combinations of white, yellow, orange, red, brown, and black are known).

### *History and Evolution*

***History:*** Starting in ancient China, various species of carp (collectively known as Asian carps) have been domesticated and reared as food fish for thousands of years. Some of these normally gray or silver species have a tendency to produce red, orange or yellow colour mutations; this was first recorded in the Jin Dynasty (265–420).

During the Tang Dynasty (618–907), it was popular to raise carp in ornamental ponds and watergardens. A natural genetic mutation produced gold (actually yellowish orange) rather than silver colouration.

People began to breed the gold variety instead of the silver variety, keeping them in ponds or other bodies of water. On special occasions at which guests were expected they would be moved to a much smaller container for display.

By the Song Dynasty (960–1279), the domestication of goldfish was firmly established. In 1162, the empress of the Song Dynasty ordered the construction of a pond to collect the red and gold variety. By this time, people outside the imperial family were forbidden to keep goldfish of the gold (yellow) variety, yellow being the imperial colour. This is probably the reason why there are more orange goldfish than yellow goldfish, even though the latter are genetically easier to breed.

During the Ming Dynasty (1368-1644), goldfish also began to be raised indoors, which led to the selection for mutations that would not be able to survive in ponds. The occurrence of other colours (apart from red and gold) was first recorded in 1276. The first occurrence of fancy-tailed goldfish was recorded in the Ming Dynasty. In 1603, goldfish were introduced to Japan, where the Ryukin and Tosakin varieties were developed. In 1611, goldfish were introduced to Portugal and from there to other parts of Europe.

During the 1620s, goldfish were highly regarded in southern Europe because of their metallic scales, and symbolized good luck and fortune. It became tradition for married men to give their wives a goldfish on their one-year anniversary, as a symbol for the prosperous years to come. This tradition quickly died, as goldfish became more available, losing their status. Goldfish were first introduced to North America around 1850 and quickly became popular in the United States.

### Related Species

Goldfish were bred from Prussian carp (*Carassius auratus gibelio*) in China, and they remain the closest wild relative of the goldfish. Previously, some sources claimed the Crucian carp (*Carassius carassius*) as the wild version of the goldfish. However, they are differentiated by several characteristics. *C. auratus* have a more pointed snout while the snout of a *C. carassius* is well rounded. *C. gibelio* often has a grey/greenish colour, while crucian carps are always golden bronze. Juvenile crucian carp have a black spot on the base of the tail which disappears with age. In C. auratus this tail spot is never present. *C. auratus* have fewer than 31 scales along the lateral line while crucian carp have 33 scales or more.

When found in nature, *C. auratus gibelio* are olive green. Introduction of goldfish into the wild can cause problems for native species. Goldfish can hybridize with certain other species of carp. Within three breeding generations, the vast majority of the hybrid spawn revert to their natural olive colour. The mutation that gave rise to the domestic goldfish is also known from other cyprinid species, such as common carp and tench. Koi may also interbreed with the goldfish to produce a sterile hybrid fish.

There are many different varieties of domesticated goldfish. Fancy goldfish are unlikely to survive in the wild because of their bright fin colours; however the hardier varieties such as the Shubunkin may survive long enough to breed with wild cousins. Common and comet goldfish can survive, and even thrive, in any climate that can support a pond.

## Biology

***Size:*** As of April 2008, the largest goldfish in the world was believed by the BBC to measure 19 inches (48 cm), and be living in the Netherlands. At the time, a goldfish named "Goldie", kept as a pet in a tank in Folkestone, England, was measured as 15 inches (38 cm) and over 2 pounds (0.91 kg), and named as the second largest in the world behind the Netherlands fish. The secretary of the Federation of British Aquatic Societies (FBAS) stated of Goldie's size that "I would think there are probably a few bigger goldfish that people don't think of as record holders, perhaps in ornamental lakes". In July 2010, a goldfish measuring 16 inches (41 cm) and 5 pounds (2.3 kg) was caught in a pond in Poole, England, thought to have been abandoned there after outgrowing a tank.

## Feeding

In the wild, the diet of goldfish consists of crustaceans, insects, and various plant matter. Like most fish, they are opportunistic feeders and do not stop eating on their own accord. Overfeeding can be deleterious to their health, typically by blocking the intestines. This happens most often with selectively bred goldfish, which have a convoluted intestinal tract. When excess food is available, they produce more waste and faeces, partly due to incomplete protein digestion. Overfeeding can sometimes be diagnosed by observing faeces trailing from the fish's cloaca.

Goldfish-specific food has less protein and more carbohydrate than conventional fish food. It is sold in two consistencies—flakes that float, and pellets that sink. Enthusiasts may supplement this diet with shelled peas (with outer skins removed), blanched green leafy vegetables, and bloodworms. Young goldfish benefit from the addition of brine shrimp to their diet. As with all animals, goldfish preferences vary.

### *Vision*

Goldfish vision is among the most studied of all vision in fishes. Goldfish have four kinds of cone cells, which are respectively sensitive to different colours: red, green, blue and ultraviolet. The ability to distinguish between four different primary colours classifies them as tetrachromats.

### *Chinese Goldfish Classification*

Chinese tradition classifies goldfish into four main types. These classifications are not commonly used in the West.

- Ce (may also be called "grass")—Goldfish without fancy anatomical features. These include the common goldfish, comet goldfish and Shubunkin.
- Wen—Goldfish have a fancy tail, e.g., Fantails and Veiltails ("Wen" is also the name of the characteristic headgrowth on such strains as Oranda and Lionhead)
- Dragon Eye—Goldfish have extended eyes, e.g., Black Moor, Bubble Eye, and Telescope Eye
- Egg—Goldfish have no dorsal fin, and usually have an 'egg-shaped' body, e.g., Lionhead (note that a Bubble Eye without a dorsal fin belongs to this group)

### *Reproduction*

Goldfish may only grow to sexual maturity with enough water and the right nutrition. Most goldfish breed in captivity, particularly in pond settings. Breeding usually happens after a significant temperature change, often in spring. Males chase gravid female goldfish (females carrying eggs), and prompt them to release their eggs by bumping and nudging them.

Goldfish, like all cyprinids, are egg-layers. Their eggs are adhesive and attach to aquatic vegetation, typically dense plants such as *Cabomba* or *Elodea* or a spawning mop. The eggs hatch within 48 to 72 hours.

Within a week or so, the fry begins to assume its final shape, although a year may pass before they develop a mature goldfish colour; until then they are a metallic brown like their wild ancestors. In their first weeks of life, the fry grow quickly—an adaptation born of the high risk of getting devoured by the adult goldfish (or other fish and insects) in their environment.

Some highly bred goldfish can no longer breed naturally due to their altered shape. The artificial breeding method called "hand stripping" can assist nature, but can harm the fish if not done correctly. In captivity, adults may also eat young that they encounter.

## Behaviour and Intelligence

***Behaviour:*** Behaviour can vary widely both because goldfish live in a variety of environments, and because their behaviour can be conditioned by their owners.

Goldfish have strong associative learning abilities, as well as social learning skills. In addition, their visual acuity allows them to distinguish between individual humans. Owners may notice that fish react favourably to them (swimming to the front of the glass, swimming rapidly around the tank, and going to the surface mouthing for food) while hiding when other people approach the tank. Over time, goldfish learn to associate their owners and other humans with food, often "begging" for food whenever their owners approach.

Goldfish are gregarious, displaying schooling behaviour, as well as displaying the same types of feeding behaviours. Goldfish may display similar behaviours when responding to their reflections in a mirror.

Goldfish that have constant visual contact with humans also stop considering them to be a threat. After being kept in a tank for several weeks, sometimes months, it becomes possible to feed a goldfish by hand without it shying away.

Goldfish have learned behaviours, both as groups and as individuals, that stem from native carp behaviour. They are a generalist species with varied feeding, breeding, and predator avoidance behaviours that contribute to their success. As fish they can be described as "friendly" towards each other. Very rarely does a goldfish harm another goldfish, nor do the males harm the females during breeding. The only real threat that goldfish present *to each other* is competing for food. Commons, comets, and other faster varieties can easily eat all the food during a feeding before fancy varieties can reach it. This can lead to stunted growth or possible starvation of fancier varieties when they are kept in a pond with their single-tailed brethren. As a result, care should be taken to combine only breeds with similar body type and swim characteristics.

### Intelligence

Goldfish have a memory-span of at least three months and can distinguish between different shapes, colours and sounds. By using positive reinforcement, goldfish can be trained to recognise and to react to light signals of different colours or to perform tricks. Fish respond to certain colours most evidently in relation to feeding. Fish learn to anticipate feedings provided they occur at around the same time every day.

### Domestication

***In Ponds:*** Goldfish are popular pond fish, since they are small, inexpensive, colourful and very hardy. In an outdoor pond or water garden, they may even survive for brief periods if ice forms on the surface, as long as there is enough oxygen remaining in the water and the pond does not freeze solid. Common goldfish, London and Bristol shubunkins, jikin, wakin, comet and some hardier fantail goldfish can be kept in a pond all year round in temperate and subtropical climates. Moor, veiltail, oranda and lionhead can be kept safely in outdoor ponds year-round only in more tropical climates and only in summer elsewhere.

Ponds small and large are fine in warmer areas (although it ought to be noted that goldfish can "overheat" in small volumes of water in summer in tropical climates). In frosty climes the depth should be at least 80 centimetres (31 in) to preclude freezing. During winter, goldfish become sluggish, stop eating and often stay on the bottom of the pond. This is normal; they become active again in the spring. Unless the pond is large enough to maintain its own ecosystem without interference from humans, a filter is important to clear waste and keep the pond clean. Plants are essential as they act as part of the filtration system, as well as a food source for the fish. Plants are further beneficial since they raise oxygen levels in the water.

Compatible fish include rudd, tench, orfe and koi, but the latter require specialised care. Ramshorn snails are helpful by eating any algae that grows in the pond. Without some form of animal population control, goldfish ponds can easily become overstocked. Fish such as orfe consume goldfish eggs.

### Mosquito Control

Like some other popular aquarium fish, such as the guppy, goldfish and other carp are frequently added to stagnant bodies of water to reduce mosquito populations. They are used to prevent the spread of West Nile Virus, which relies on mosquitoes to migrate. However, introducing goldfish has often had negative consequences for local ecosystems.

## Mantis Shrimp

Mantis shrimp or stomatopods are marine crustaceans, the members of the order Stomatopoda. They may reach 30 centimetres (12 in) in length, though in exceptional cases have been recorded at up to 38 cm (15 in).

The carapace of mantis shrimp covers only the rear part of the head and the first four segments of the thorax. Mantis shrimp appear in a variety of colours, from shades of browns to bright neon colours.

Although they are common animals and among the most important predators in many shallow, tropical and sub-tropical marine habitats, they are poorly understood as many species spend most of their life tucked away in burrows and holes.

Called "sea locusts" by ancient Assyrians, "prawn killers" in Australia and now sometimes referred to as "thumb splitters" – because of the animal's ability to inflict painful gashes if handled incautiously – mantis shrimp sport powerful claws that they use to attack and kill prey by spearing, stunning, or dismemberment.

Although it only happens rarely, some larger species of mantis shrimp are capable of breaking through aquarium glass with a single strike from this weapon.

### *Ecology*

These aggressive and typically solitary sea creatures spend most of their time hiding in rock formations or burrowing intricate passageways in the sea bed. They either wait for prey to chance upon them or, unlike most crustaceans, at times they hunt, chase, and kill prey. They rarely exit their homes except to feed and relocate, and can be diurnal, nocturnal, or crepuscular, depending on the species. Most species live in tropical and subtropical seas (Indian and Pacific Oceans between eastern Africa and Hawaii), although some live in temperate seas.

### *Classification and the Claw*

Around 400 species of mantis shrimp have currently been described worldwide; all living species are in the suborder Unipeltata.

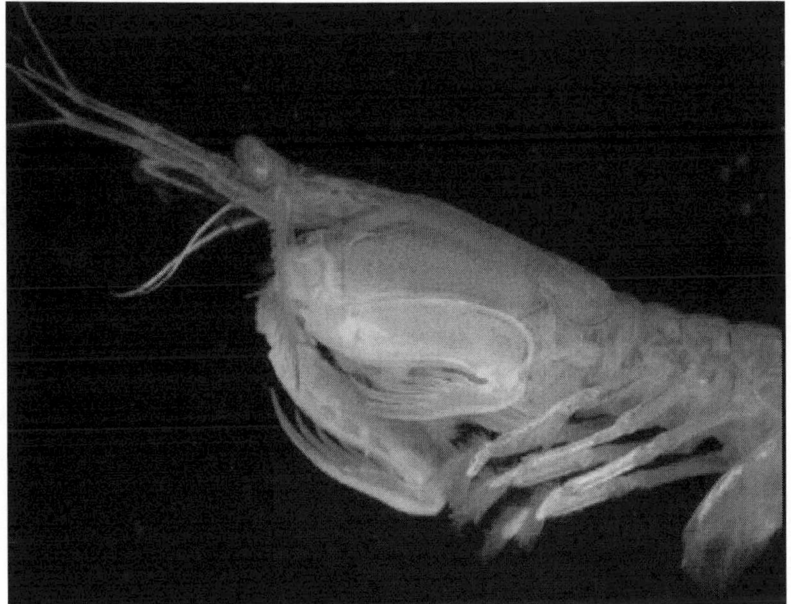

***Figure:*** *Squilla mantis, showing the spearing appendages*

***Figure:*** *Mantis shrimp from the front*

They are commonly separated into two distinct groups determined by the manner of claws they possess:

- *Spearers* are armed with spiny appendages topped with barbed tips, used to stab and snag prey.
- *Smashers*, on the other hand, possess a much more developed club and a more rudimentary spear (which is nevertheless quite sharp and still used in fights between their own kind); the club is used to bludgeon and smash their meals apart. The inner aspect of the dactyl (the terminal portion of the appendage) can also possess a sharp edge, with which the animal can cut prey while it swims.

Both types strike by rapidly unfolding and swinging their raptorial claws at the prey, and are capable of inflicting serious damage on victims significantly greater in size than themselves. In smashers, these two weapons are employed with blinding quickness, with an acceleration of 10,400 *g* (102,000 $m/s^2$ or 335,000 $ft/s^2$) and speeds of 23 m/s from a standing start, about the acceleration of a .22 calibre bullet. Because they strike so rapidly, they generate cavitation bubbles between the appendage and the striking surface. The collapse of these cavitation bubbles produces measurable forces on their prey in addition to the instantaneous forces of 1,500 newtons that are caused by the impact of the appendage against the striking surface, which means that the prey is hit twice by a single strike; first by the claw and then by the collapsing cavitation bubbles that immediately follow. Even if the initial strike misses the prey, the resulting shock wave can be enough to stun or kill the prey.

The snap can also produce sonoluminescence from the collapsing bubble. This will produce a very small amount of light and high temperatures in the range of several thousand kelvins within the collapsing bubble, although both the light and high temperatures are too weak and short-lived to be detected without advanced scientific equipment. The light emission and temperature increase probably have no biological significance but are rather side-effects of the rapid snapping motion. Pistol shrimp produce this effect in a very similar manner.

Smashers use this ability to attack snails, crabs, molluscs and rock oysters, their blunt clubs enabling them to crack the shells of their prey into pieces. Spearers, on the other hand, prefer the meat of softer animals, like fish, which their barbed claws can more easily slice and snag.

### *Eyes*

The midband region of the mantis shrimp's eye is made up of six rows of specialised ommatidia. Four rows carry 16 differing sorts of photoreceptor pigments, 12 for colour sensitivity, others for colour filtering. The mantis shrimp has such good eyes it can perceive both polarised light and multispectral images. Their eyes (both mounted on mobile stalks and constantly moving about independently of each other) are similarly variably coloured and are considered to be the most complex eyes in the animal kingdom. They permit both serial and parallel analysis of visual stimuli.

Each compound eye is made up of up to 10,000 separate ommatidia of the apposition type. Each eye consists of two flattened hemispheres separated by six parallel rows of highly specialised ommatidia, collectively called the midband, which divides the eye into three regions. This is a configuration that makes it possible for mantis shrimp to see objects with three different parts of the same eye. In other words, each individual eye possesses trinocular vision and depth perception. The upper and lower hemispheres are used primarily for recognition of forms and motion, not colour vision, like the eyes of many other crustaceans.

Rows 1–4 of the midband are specialised for colour vision, from ultra-violet to longer wavelengths, but aren't currently believed to be sensitive to infrared light. The optical elements in these rows have eight different classes of visual pigments and the rhabdom is divided into three different pigmented layers (tiers), each for different wavelengths. The three tiers in rows 2 and 3 are separated by colour filters (intrarhabdomal filters) that can be divided into four distinct classes, two classes in each row. It is organised like a sandwich; a tier, a colour filter of one class, a tier again, a colour filter of another class, and then a last tier. Rows 5–6 are segregated into different tiers too, but have only one class of visual pigment (a ninth class) and are specialised for polarisation vision. They can detect different planes of polarised light. A tenth class of visual pigment is found in the dorsal and ventral hemispheres of the eye.

The midband only covers a small area of about 5°–10° of the visual field at any given instant, but like in most crustaceans, the eyes are mounted on stalks. In mantis shrimps the movement of the stalked eye is unusually free, and can be driven in all possible axes, up to at least 70°, of movement by eight individual eyecup muscles divided into six functional groups. By using these muscles to scan the surroundings with the midband, they can add information about forms, shapes and landscape which cannot be detected by the upper and lower hemisphere of the eye. They can also track moving objects using large, rapid eye movements where the two eyes move independently. By combining different techniques, including saccadic movements, the midband can cover a very wide range of the visual field.

Some species have at least 16 different photoreceptor types, which are divided into four classes (their spectral sensitivity is further tuned by colour filters in the retinas), 12 of them for colour analysis in the different wavelengths (including four which are sensitive to ultraviolet light) and four of them for analysing polarised light. By comparison, most humans have only four visual pigments, three dedicated to see colour but the lenses block ultraviolet light. The visual information leaving the retina seems to be processed into numerous parallel data streams leading into the central nervous system, greatly reducing the analytical requirements at higher levels.

At least two species have been reported to be able to detect circularly polarised light, and in some cases their biological quarter-wave plates perform more uniformly over the entire visual spectrum than any current man-made polarizing optics, the application of which it is speculated could be applied to a new type of optical media that performs even better than the current generation of Blu-ray disc technology.

The species *Gonodactylus smithii* is the only organism known to simultaneously detect the four linear and two circular polarization components required for Stokes parameters, which yield a full description of polarization. It is thus believed to have optimal polarization vision.

### *Suggested Advantages of Visual System*

It is not clear what advantage sensitivity to polarization confers; however polarization vision is used by other animals for sexual signalling and secret communication that avoids the attention of predators. This mechanism could provide an evolutionary advantage; it only requires small changes to the cell in the eye and would be easily selected for. The eyes of mantis shrimp may enable them to recognise different types of coral, prey species (which are often transparent or semi-transparent), or predators, such as barracuda, which have shimmering scales. Alternatively, the manner in which mantis shrimp hunt (very rapid movements of the claws) may require very accurate ranging information, which would require accurate depth perception.

The fact that those with the most advanced vision also are the species with the most colourful bodies suggests the evolution of colour vision has taken the same direction as the peacock's tail. During mating rituals, mantis shrimp actively fluoresce, and the wavelength of this fluorescence matches the wavelengths detected by their eye pigments. Females are only fertile during certain phases of the tidal cycle; the ability to perceive the phase of the moon may therefore help prevent wasted mating efforts. It may also give mantis shrimp information about the size of the tide, which is important for species living in shallow water near the shore.

### Behaviour

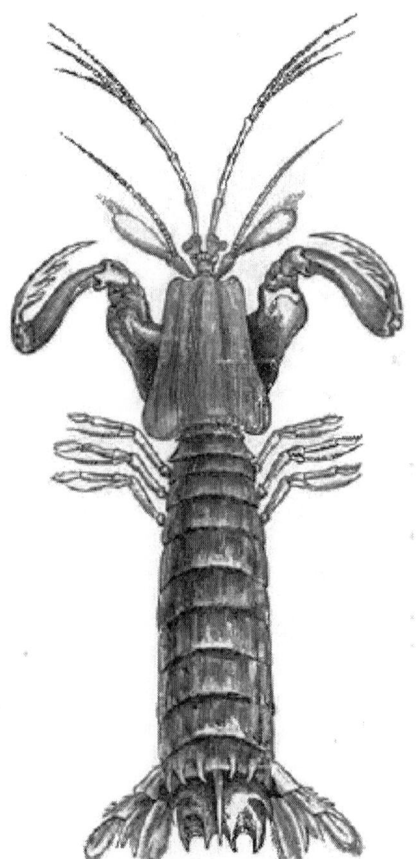

***Figure:*** *An 1896 drawing of a mantis shrimp*

Mantis shrimp are long-lived and exhibit complex behaviour, such as ritualised fighting. Some species use fluorescent patterns on their bodies for signalling with their own and maybe even other species, expanding their range of behavioural signals. They can learn and remember well, and are able to recognise individual neighbours with whom they frequently interact. They can recognise them by visual signs and even by individual smell. Many have developed complex social behaviour to defend their space from rivals.

In a lifetime, they can have as many as 20 or 30 breeding episodes. Depending on the species, the eggs can be laid and kept in a burrow, or they can be carried around under the female's tail until they hatch. Also depending on the species, male and female may come together only to mate, or they may bond in monogamous long-term relationships.

In the monogamous species, the mantis shrimp remain with the same partner for up to 20 years. They share the same burrow and may be able to coordinate their activities. Both sexes often take care of the eggs (biparental care). In *Pullosquilla* and some species in *Nannosquilla,* the female will lay two clutches of eggs: one that the male tends and one that the female tends. In other species, the female will look after the eggs while the male hunts for both of them. Once the eggs hatch, the offspring may spend up to three months as plankton.

Although stomatopods typically display the standard locomotion types as seen in true shrimp and lobsters, one species, *Nannosquilla decemspinosa,* has been observed flipping itself into a crude wheel. The species lives in shallow, sandy areas. At low tides, *N. decemspinosa* is often stranded by its short rear legs, which are sufficient for locomotion when the body is supported by water, but not on dry land. The mantis shrimp then performs a forward flip in an attempt to roll towards the next tide pool. *N. decemspinosa* has been observed to roll repeatedly for 2 metres (6.6 ft), but specimens typically travel less than 1 m (3.3 ft).

### *Culinary Uses*

In Japanese cuisine, the mantis shrimp is eaten boiled as a sushi topping, and occasionally, raw as sashimi; and is called *shako.* Mantis shrimp are abundant in the coastal regions of south Vietnam, known in Vietnamese as *tôm tít* or *tôm tích.* The shrimp can be steamed, boiled, grilled or dried; used with pepper + salt + lime, fish sauce + tamarind or fennel.

In Cantonese cuisine, the mantis shrimp is known as "pissing shrimp" (Mandarin pinyin: *lài niào xiâ,* modern Cantonese: *laaih niu hâ*) because of their tendency to shoot a jet of water when picked up. After cooking, their flesh is closer to that of lobsters than that of shrimp, and like lobsters, their shells are quite hard and require some pressure to crack. Usually they are deep fried with garlic and chili peppers. In the Mediterranean countries the mantis shrimp *Squilla mantis* is a common seafood, especially on the Adriatic coasts (canocchia) and the Gulf of Cádiz (galera).

In the Philippines, the mantis shrimp is known as tatampal, hipong-dapa or alupihang-dagat and is cooked and eaten like shrimp. The usual concerns associated with consuming seafood are an issue with mantis shrimp when those dwell in contaminated waters. In Hawaii, some have grown unusually large in the very dirty waters of the Grand Ala Wai Canal in Waikiki.

### Aquaria

***Figure:*** *A colourful stomatopod, the peacock mantis shrimp, (*Odontodactylus scyllarus*) seen in the Andaman Sea off Thailand*

Some saltwater aquarists keep stomatopods in captivity. The peacock mantis is especially colourful and desired in the trade.

While some aquarists value mantis shrimp, others consider them harmful pests, because they:

- Are voracious predators, eating other desirable inhabitants of the tank,
- Can, in some of the largest species, break aquarium glass by striking it

- In some rock-burrowing species, can do more damage to live rock than the fishkeeper would prefer

The live rock with mantis shrimp burrows are actually considered useful by some in the marine aquarium trade and are often collected. It is not uncommon for a piece of live rock to convey a live mantis shrimp into an aquarium.

Once inside the tank, they may feed on fish, and other inhabitants. They are notoriously difficult to catch when established in a well-stocked tank, and there are accounts of them breaking glass tanks. It should be noted that while stomatopods do not eat coral, the smashers can damage it if they wish to make a home within it.

## Sea Urchins Ocean Hedgehogs

Sea urchins or urchins, sometimes called sea hedgehogs, are small, spiny, globular animals which, with their close kin, such as sand dollars, constitute the class Echinoidea of the echinoderm phylum. There are c. 950 species of echinoids inhabiting all oceans from the intertidal to 5000 metres deep. Their shell, or "test", is round and spiny, typically from 3 to 10 cm (1.2 to 3.9 in) across. Common colours include black and dull shades of green, olive, brown, purple, blue, and red. They move slowly, feeding mostly on algae. Sea otters, wolf eels, triggerfish, and other predators feed on them. Their roe is a delicacy in many cuisines.

The name "urchin" is an old name for the round spiny hedgehogs that sea urchins resemble.

### *Taxonomy*

Sea urchins are members of the phylum Echinodermata, which also includes sea stars, sea cucumbers, brittle stars, and crinoids. Like other echinoderms, they have fivefold symmetry (called pentamerism) and move by means of hundreds of tiny, transparent, adhesive "tube feet". The symmetry is not obvious in the living animal, but is easily visible in the dried test. *Echinodermate* means "spiny skin" in Greek.

Specifically, the term "sea urchin" refers to the "regular echinoids", which are symmetrical and globular. The term includes several different taxonomic groups: the order Echinoida, the order Cidaroida or "slate-pencil urchins", which have very thick, blunt spines, and others. Besides sea urchins, the class Echinoidea also includes three groups of "irregular" echinoids: flattened sand dollars, sea biscuits, and heart urchins.

Together with sea cucumbers (Holothuroidea), they make up the subphylum Echinozoa, which is characterized by a globoid shape without arms or projecting rays. Sea cucumbers and the irregular echinoids have secondarily evolved diverse shapes. Although many sea cucumbers have branched tentacles surrounding the oral opening, these have originated from modified tube feet and are not homologous to the arms of the crinoids, sea stars, and brittle stars.

### Anatomy

Urchins typically range in size from 6 to 12 cm (2.4 to 4.7 in), although the largest species can reach up to 36 cm (14 in).

### Fivefold Symmetry

Like other echinoderms, sea urchins are bilaterans. Their early larvae have bilateral symmetry, but they develop fivefold symmetry as they mature. This is most apparent in the "regular" sea urchins, which have roughly spherical bodies, with five equally sized parts radiating out from their central axes.

Several sea urchins, however, including the sand dollars, are oval in shape, with distinct front and rear ends, giving them a degree of bilateral symmetry. In these urchins, the upper surface of the body is slightly domed, but the underside is flat, while the sides are devoid of tube feet. This "irregular" body form has evolved to allow the animals to burrow through sand or other soft materials.

***Organs and Test:*** The lower half of a sea urchin's body is referred to as the oral surface, because it contains the mouth, while the upper half is the aboral surface. The internal organs are enclosed in a hard test or shell composed of fused plates of calcium carbonate covered by a thin dermis and epidermis. The test is rigid, and divides into five ambulacral grooves separated by five interambulacral areas. Each of these areas consists of two rows of plates, so the test includes 20 rows in total. The plates are covered in rounded tubercles, to which the spines are attached. The inner surface of the test is lined by peritoneum.

### Feet

Sea urchins have tube feet, which arise from the five ambulacral grooves. Tube feet are moved by a water vascular system. This water vascular system works through hydraulic pressure, allowing the Sea Urchin to pump water into and out of the tube feet, enabling it to locomote.

### Mouth/Anus

The mouth lies in the centre of the oral surface in regular urchins, or towards one end in irregular urchins. It is surrounded by lips of softer tissue, with numerous small, bony pieces embedded in it. This area, called the peristome, also includes five pairs of modified tube feet and, in many species, five pairs of gills. On the upper surface, opposite the mouth, is a region termed the periproct, which surrounds the anus. The periproct contains a variable number of hard plates, depending on species, one of which contains the madreporite.

### Endoskeleton

The sea urchin builds its spicules, the sharp crystalline "bones" that constitute the animal's endoskeleton, in the larval stage. The fully formed spicule is composed of a single crystal with an unusual morphology. It has no facets, and within 48 hours of fertilization assumes a shape that looks very much like the Mercedes-Benz logo.

In other echinoderms, the endoskeleton is associated with a layer of muscle that allows the animal to move its arms or other body parts. This is entirely absent in sea urchins, which are unable to move in this way.

### Spines

The spines, long and sharp in some species, protect the urchin from predators. They inflict a painful wound when they penetrate human skin, but are not dangerous if fully removed promptly, if left there may be further problems. It is not clear if the spines are venomous (unlike the pedicellariae between the spines, which are venomous).

Typical sea urchins have spines that are 1 to 3 cm (0.39 to 1.2 in) in length, 1 to 2 mm (0.039 to 0.079 in) thick, and not terribly sharp. *Diadema antillarum*, familiar in the Caribbean, has thin, potentially dangerous spines that can reach 10 to 30 cm (3.9 to 12 in) long.

### Reproductive Organs

Sea urchins are dioecious, having separate male and female sexes, although distinguishing the two is not easy, except for their locations on the sea bottom. Males generally choose an elevated and exposed location, so their milt can be broadcast by sea currents. Females generally choose a low-lying location in sea bottom crevices, presumably so the tiny larvae can have better protection from predators. Indeed, very small sea urchins are found hiding beneath rocks. Regular sea urchins have five gonads, lying underneath the interambulacral regions of the test, while the irregular forms have only four, with the hindmost gonad being absent. Each gonad has a single duct rising from the upper pole to open at a gonopore lying in one of the genital plates surrounding the anus. The gonads are lined with muscles underneath the peritoneum, and these allow the animal to squeeze its gametes through the duct and into the surrounding sea water where fertilization takes place.

### Physiology

***Digestion:*** The mouth of most sea urchins is made up of five calcium carbonate teeth or jaws, with a fleshy, tongue-like structure within. The entire chewing organ was known as Aristotle's lantern (image), from Aristotle's description in his *History of Animals*:

> *...the urchin has what we mainly call its head and mouth down below, and a place for the issue of the residuum up above. The urchin has, also, five hollow teeth inside, and in the middle of these teeth a fleshy substance serving the office of a tongue. Next to this comes the esophagus, and then the stomach, divided into five parts, and filled with excretion, all the five parts uniting at the anal vent, where the shell is perforated for an outlet... In reality the mouth-apparatus of the urchin is continuous from one end to the other, but to outward appearance it is not so, but looks like a horn lantern with the panes of horn left out. (Tr. D'Arcy Thompson)*

However, this has recently been proven to be a mistranslation. Aristotle's lantern is actually referring to the whole shape of sea urchins, which look like the ancient lamps of Aristotle's time.

Recent research has shown the sea urchin's teeth are self-sharpening; it can chew through stone. Heart urchins are unusual in not having a lantern. Instead, the mouth is surrounded by cilia that pull strings of mucus-containing food particles towards a series of grooves around the mouth.

The lantern, where present, surrounds both the mouth cavity and the pharynx. At the top of the lantern, the pharynx opens into the esophagus, which runs back down the outside of the lantern, to join the small intestine and a single caecum. The small intestine runs in a full circle around the inside of the test, before joining the large intestine, which completes another circuit in the opposite direction.

From the large intestine, a rectum ascends towards the anus. Despite the names, the small and large intestines of sea urchins are in no way homologous to the similarly named structures in vertebrates. Digestion occurs in the intestine, with the caecum producing further digestive enzymes.

An additional tube, called the siphon, runs beside much of the intestine, opening into it at both ends. It may be involved in resorption of water from food.

### Circulation

Sea urchins possess both a water vascular system and a hemal system, the latter containing blood. However, the main circulatory fluid fills the general body cavity, or coelom. This fluid contains phagocytic coelomocytes, which move through the vascular and hemal systems. The coelomocytes are an essential part of blood clotting, but also collect waste products and actively remove them from the body through the gills and tube feet.

### Respiration

Most sea urchins possess five pairs of external gills, located around the mouth. These thin-walled projections of the body cavity are the main organs of respiration in those urchins that possess them. Fluid can be pumped through the gills' interiors by muscles associated with the lantern, but this is not continuous, and occurs only when the animal is low on oxygen. Tube feet can also act as respiratory organs, and are the primary sites of gas exchange in heart urchins and sand dollars, both of which lack gills.

### Nervous System

The nervous system of sea urchins has a relatively simple layout. There is no true brain. The centre is a large nerve ring encircling the mouth just inside the lantern. From the nerve ring, five nerves radiate underneath the radial canals of the water vascular system, and branch into numerous finer nerves to innervate the tube feet, spines, and pedicellariae.

### Senses

Sea urchins are sensitive to touch, light, and chemicals. Although they do not have eyes or eye spots, recent research suggests their entire body might function as one compound eye. They also have statocysts, called *spheridia*, located within the ambulacral plates to help the animal remain upright.

### *Development*

Ingression of primary mesenchyme cells:

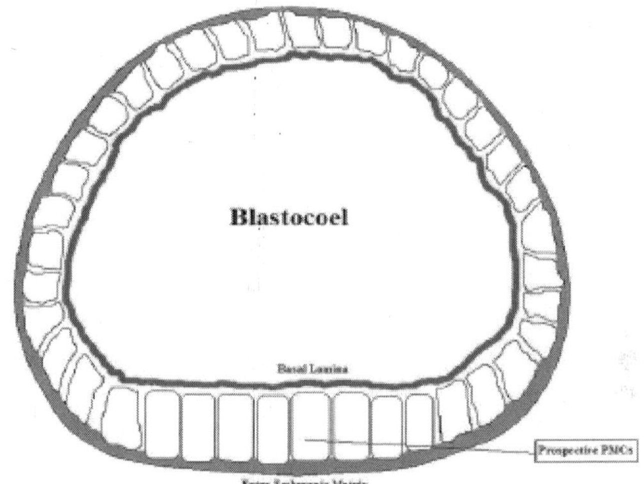


***Figure:*** *Sea urchin blastula*

During early development, the sea urchin embryo undergoes 10 cycles of cell division, resulting in a single epithelial layer enveloping a blastocoel. The embryo must then begin gastrulation, a multipart process which involves the dramatic rearrangement and invagination of cells to produce the three germ layers.

The first step of gastrulation is the epithelial-to-mesenchymal transition and ingression of primary mesenchyme cells into the blastocoel. Primary mesenchyme cells, or PMCs, are located in the vegetal plate specified to become mesoderm. Prior to ingression, PMCs exhibit all the features of other epithelial cells that comprise the embryo. Cells of the epithelium are bound basally to a laminal matrix and apically to an extraembryonic matrix. The apical microvilli of these cells reach into the hyaline layer, a component of the extraembryonic matrix. Neighbouring epithelial cells are also connected to each other through apical junctions, protein complexes containing adhesion molecules, such as cadherins, linked to catenins.

As PMCs begin to undergo an epithelial-to-mesenchymal transition, the lamina which binds them dissolves to begin the mechanical release of the cells. Expression of the membrane protein that binds laminin, integrin, also becomes irregular at the beginning of ingression. The microvilli which secure PMCs to the hyaline layer shorten, as the cells reduce their affinity for the extraembryonic matrix. These cells concurrently increase their affinity for other components of the basal matrix, such as fibronectin, in part driving the movement of cells inward. The apical junctions which bind PMCs to their neighbouring epithelial cells become disrupted during this transition, and are absent in cells that have fully ingressed into the blastocoel. Because staining for cadherins and catenins in ingressing cells decreases and develops as intracellular accumulations, apical junctions are thought to be cleared by endocytosis during ingression.

Once the PMCs disrupt all attachment to their former location, the cells themselves change their morphology by contracting their apical surfaces, apical constriction, and enlarging their basal surfaces, thus acquiring a "bottle cell" phenotype. Cytoskeletal rearrangements mediate the shape changes of PMCs; though the cytoskeleton assists in the mechanics of ingression, other mechanisms drive the process. Experimentally disrupting microtubule dynamics in the species *Strongylocentrotus pupuratus* by applying colchicine stalls the ingression of PMCs, but does not inhibit it. Similarly, experimentally disrupting actin-myosin contraction using inhibitors slows down ingression, but does not arrest the process.

The morphogenetic movements of the PMCs are an autonomous cellular behaviour. Experimentally grafting PMCs into heterotopic tissue does not prevent the cells from ingressing. In studies where PMCs are cultured in insolation, the cells were observed to gain affinity for fibronectin and simultaneously lose affinity for extraembryonic matrix, independent of the embryonic environment.

### Life History

At first glance, sea urchins often appear sessile, i.e. incapable of moving. Sometimes, the most visible life sign is the spines, which attach to ball-and-socket joints and can point in any direction. In most urchins, touch elicits a prompt reaction from the spines, which converge toward the touch point. Sea urchins have no visible eyes, legs, or means of propulsion, but can move freely over hard surfaces using adhesive tube feet, working in conjunction with the spines.

### Reproduction

In most cases, the female Sea Urchin's eggs float freely in the sea, but some species hold onto them with their spines, affording them a greater degree of protection. The fertilized egg, once met with the free floating sperm released by males, develops into a free-swimming blastula embryo in as little as 12 hours. Initially a simple ball of cells, the blastula soon transforms into a cone-shaped echinopluteus larva. In most species, this larva has 12 elongated arms. The arms are lined with bands of cilia that capture food particles and transport them to the mouth. In a few species, the blastula contains supplies of nutrient yolk and lacks arms, since it has no need to feed.

It may take several months for the larva to complete its development, which begins with the formation of the test plates around the mouth and anus. Soon the larva sinks to the bottom and metamorphoses into adult form in as little as one hour. In some species, adults reach their maximum size in about five years.

### *Ecology*

Sea urchins feed mainly on algae, but can also feed on sea cucumbers and a wide range of invertebrates, such as mussels, polychaetes, sponges, brittle stars and crinoids. Population densities vary by habitat, with more dense populations being found in barren areas as compared to kelp stands.

Even in these barren areas, greatest densities are also found in shallow water. Populations are also generally found in deeper water if wave action is present. Densities also decrease in winter when storms cause them to seek protection in cracks and around larger underwater structures. The shingle urchin (*Colobocentrotus atratus*), which lives on exposed shorelines, is particularly resistant to wave action.

Sea urchins are some of the favourite foods of sea otters, and are also the main source of nutrition for wolf eels. Left unchecked, urchins devastate their environments, creating what biologists call an urchin barren, devoid of macroalgae and associated fauna. Sea otters have re-entered British Columbia, dramatically improving coastal ecosystem health.

### *Evolutionary History*

The earliest echinoid fossils date to the upper part of the Ordovician period (*circa* 450 MYA), and the taxon has survived to the present as a successful and diverse group of organisms. Spines may be present in well-preserved specimens, but usually only the test remains. Isolated spines are common as fossils. Some echinoids (such as *Tylocidaris clavigera*, from the Cretaceous period's English Chalk Formation) had very heavy, club-shaped spines that would be difficult for an attacking predator to break through and make the echinoid awkward to handle. Such spines simplify walking on the soft sea floor.

Most of the fossil echinoids from the Paleozoic era are incomplete, consisting of isolated spines and small clusters of scattered plates from crushed individuals, mostly in Devonian and Carboniferous rocks. The shallow-water limestones from the Ordovician and Silurian periods of Estonia are famous for echinoids. Paleozoic echinoids probably inhabited relatively quiet waters. Because of their thin tests, they would certainly not have survived in the wave-battered coastal waters inhabited by many modern echinoids. During the upper part of the Carboniferous period, a marked decline in echinoid diversity occurred, and this trend continued to the Permian period. They neared extinction at the end of the Paleozoic era, with just six species known from the Permian period. Only two lineages survived this period's massive extinction and into the Triassic: the genus *Miocidaris*, which gave rise to modern cidaroida (pencil urchins), and the ancestor that gave rise to the euechinoids. By the upper part of the Triassic period, their numbers began to increase again. Cidaroids have changed very little since the Late Triassic and are today considered to be living fossils.

The euechinoids, on the other hand, diversified into new lineages throughout the Jurassic and into the Cretaceous periods, and from them emerged the first irregular echinoids (superorder Atelostomata) during the early Jurassic, and later the other superorder (Gnathostomata) of irregular urchins, which evolved independently. These superorders today represent 47% of all extant species of echinoids because of their adaptive breakthroughs, which allowed them to exploit habitats and food sources unavailable to regular echinoids. During the Mesozoic and Cenozoic eras, the echinoids flourished. Most echinoid fossils are often abundant in the restricted localities and formations where they occur. An example of this is *Enallaster*, which exists by the thousands in certain outcrops of limestone from the Cretaceous period in Texas. Many fossils of the Late Jurassic *Plesiocidaris* still have the spines attached.

Some echinoids, such as *Micraster*, which is found in the Cretaceous period Chalk Formation of England and France, serve as zone or index fossils. Because they evolved rapidly, they aid geologists in dating the surrounding rocks. However, most echinoids are not abundant enough and are of too limited range to serve as zone fossils.

In the early Tertiary (*circa* 65 to 1.8 MYA), sand dollars (order Clypeasteroida) arose. Their distinctive, flattened tests and tiny spines were adapted to life on or under loose sand. They form the newest branch on the echinoid tree.

## Relation to Humans

***In Biology:*** Sea urchins are traditional model organisms in developmental biology. This use originated in the 1800s, when their embryonic development became easily viewed by microscopy. It was the transparency of the urchin's eggs that enabled them to be used to observe that sperm cells actually fertilize ova.

The recent sequencing of the sea urchin genome established homology between sea urchin and vertebrate immune system-related genes. Sea urchins code for at least 222 toll-like receptor genes and over 200 genes related to the nod-like-receptor family found in vertebrates. This increases its usefulness as a valuable model organism for studying the evolution of innate immunity.

## As Food

The gonads of both male and female sea urchins, usually called sea urchin roe or corals, are culinary delicacies in many parts of the world. In cuisines around the Mediterranean, *Paracentrotus lividus* is often eaten raw, with lemon., and known as *ricci* on Italian menus where it is sometimes used in pasta sauces. It can also flavour omelettes, scrambled eggs, fish soup, mayonnaise, béchamel sauce for tartlets, the *boullie* for a soufflé, or Hollandaise sauce to make a fish sauce. In Chilean cuisine, it is served raw with lemon, onions, and olive oil.

Though the edible *Strongylocentrotus droebachiensis* is found in the North Atlantic, it is not widely eaten. However, sea urchins (called *uutuk* in Alutiiq) are commonly eaten by the Alaska Native population around Kodiak Island. It is commonly exported, mostly to Japan.

In the West Indies, slate pencil urchins are eaten.

On the Pacific Coast of North America, *Strongylocentrotus franciscanus* was praised by Euell Gibbons; *Strongylocentrotus purpuratus* is also eaten.

In New Zealand, *Evechinus chloroticus*, known as *kina* in Maori, is a delicacy, traditionally eaten raw. Though New Zealand fishermen would like to export them to Japan, their quality is too variable.

In Japan, sea urchin is known as *uni*, and its roe can retail for as much as A$450/kg; it is served raw as *sashimi* or in *sushi*, with soy sauce and *wasabi*. Japan imports large quantities from the United States, South Korea, and other producers. Japanese demand for sea urchin corals has raised concerns about overfishing.

Native Americans in California are also known to eat sea urchins.

### *Aquaria*

Some species of sea urchins, such as the slate pencil urchin (*Eucidaris tribuloides*), are commonly sold in aquarium stores. Some species are effective at controlling hair algae, and they make good additions to an invertebrate tank.

## Fish and Chips

Fish and chips is a take-away food that consists of battered fish, commonly cod or haddock, and deep-fried chips, sometimes accompanied by mushy peas. The dish originated in the United Kingdom in the 19th century.

### *History*

Fish and chips became a stock meal among the working classes in the United Kingdom as a consequence of the rapid development of trawl fishing in the North Sea, and the development of railways which connected the ports to major industrial cities during the second half of the 19th century, which meant that fresh fish could be rapidly transported to the heavily populated areas. Deep-fried fish was first introduced into Britain during the 17th century by Jewish refugees from Portugal and Spain, and is derived from pescado frito. In 1860, the first fish and chip shop was opened in London by Joseph Malin.

Deep-fried chips (slices or pieces of potato) as a dish may have first appeared in Britain in about the same period: the *Oxford English Dictionary* notes as its earliest usage of "chips" in this sense the mention in Dickens' *A Tale of Two Cities* (published in 1859): "Husky chips of potatoes, fried with some reluctant drops of oil".

The modern fish-and-chip shop ("chippy" or "chipper" in modern British slang) originated in the United Kingdom, although outlets selling fried food occurred commonly throughout Europe. Early fish-and-chip shops had only very basic facilities. Usually these consisted principally of a large cauldron of cooking fat, heated by a coal fire. During World War II fish and chips remained one of the few foods in the United Kingdom not subject to rationing.

In the United Kingdom and Ireland, the Fish Labelling Regulations 2003 enact directive 2065/2001/EC and generally means that "fish" must be sold with the particular species named; so "cod and chips" now appears on menus rather than the more vague "fish and chips". In the United Kingdom the Food Standards Agency guidance excludes caterers from this; but several local Trading Standards authorities and others do say it cannot be sold merely as "fish and chips".

### England

The dish became popular in wider circles in London and South East England in the middle of the 19th century (Charles Dickens mentions a "fried fish warehouse" in *Oliver Twist*, first published in 1838), while in the north of England a trade in deep-fried chipped potatoes developed. The first chip shop stood on the present site of Oldham's Tommyfield Market. It remains unclear exactly when and where these two trades combined to become the fish-and-chip shop industry we know. A Jewish immigrant, Joseph Malin opened the first recorded combined fish-and-chip shop in London in 1860 or in 1865; a Mr Lees pioneered the concept in the North of England, in Mossley, in 1863.

The concept of a fish restaurant was introduced by Samuel Isaacs (born 1856 in Whitechapel, London; died 1939 in Brighton, Sussex) who ran a thriving wholesale and retail fish business throughout London and the South of England in the latter part of the 19th century. Isaacs' first restaurant opened in London in 1896 serving fish and chips, bread and butter, and tea for nine pence, and its popularity ensured a rapid expansion of the chain.

The restaurants were carpeted, had waited service, tablecloths, flowers, china and cutlery, and made the trappings of upmarket dining affordable to the working classes for the first time. They were located in Tottenham Court Road, St Pancras, The Strand, Hoxton, Shoreditch, Brixton and other London districts, as well as Clacton, Brighton, Ramsgate, Margate and other seaside resorts in southern England. Menus were expanded in the early 20th century to include meat dishes and other variations as their popularity grew to a total of thirty restaurants. Sam Isaacs' trademark was the phrase "This is the Plaice" combined with a picture of the punned-upon fish in question. A glimpse of the old Brighton restaurant at No.1 Marine Parade can be seen in the background of Norman Wisdom's 1955 film *One Good Turn* just as Norman/Pitkin runs onto the seafront; this is now the site of a Harry Ramsden's fish and chips restaurant. A blue plaque at Oldham's Tommyfield Market marks the first chips fried in Britain in 1860, and the origin of the fish and chip shop and fast food industries in Britain.

### Scotland

Dundee City Council claims that "...in the 1870s, that glory of British gastronomy - the chip - was first sold by Belgian immigrant Edward De Gernier in the city's Greenmarket."

In Edinburgh, a combination of Gold Star brown sauce and water or malt vinegar, known as "sauce", or more specifically as "chippy sauce", has great popularity.

### Ireland

In Ireland, the first fish and chips were sold by an Italian immigrant, Giuseppe Cervi, who mistakenly stepped off an America-bound ship at Cobh (then called Queenstown) in County Cork and walked all the way to Dublin. He started by selling fish and chips outside Dublin pubs from a handcart. He then found a permanent spot in Great Brunswick Street (now Pearse Street). His wife Palma would ask customers "Uno di questa, uno di quella?" This phrase (meaning "one of this, one of the other") entered the vernacular in Dublin as "one and one", which is still a way of referring to fish and chips in the city.

### Composition

***Cooking:*** Traditional frying uses beef dripping or lard; however, vegetable oils, such as peanut oil (used because of its relatively high smoke point) now predominate. A minority of vendors in the north of England and Scotland and the majority of vendors in Northern Ireland still use dripping or lard, as it imparts a different flavour to the dish, but it has the side effect of making the fried chips unsuitable for vegetarians and for adherents of certain faiths. Lard is used in some living industrial history museums, such as the Black Country Living Museum.

### Thickness

British chips are traditionally thicker than American-style French fries sold by major multinational fast food chains, resulting in a lower fat content per portion. In their homes or in non-chain restaurants, people in or from the United States may eat a thick type of chip, more similar to the British variant, sometimes referred to as steak fries.

How much cooking fat soaks into the potato depends on the surface area and how long they are cooked. Chips have a smaller surface area per unit weight than French fries, which means absorbing less oil in a given time. On the other hand, chips, being thicker, take longer to cook than fries.

### Batter

UK chippies traditionally use a simple water and flour batter, adding a little sodium bicarbonate (baking soda) and a little vinegar to create lightness, as they create bubbles in the batter. Other recipes may use beer or milk batter, where these liquids are often substitutes for water. The carbon dioxide in the beer lends a lighter texture to the batter. Beer also results in an orange-brown colour. A simple beer batter might consist of a 2:3 ratio of flour to beer by volume. The type of beer makes the batter taste different: some prefer lager whereas others use stout or bitter.

### *Choice of Fish*

In Britain and Ireland, cod and haddock appear most commonly as the fish used for fish and chips, but vendors also sell many other kinds of fish, especially other white fish, such as pollock or coley, plaice, skate, and ray (particularly popular in Ireland); and huss or rock salmon (a term covering several species of dogfish and similar fish). In Northern Ireland, cod, plaice or whiting appear most commonly in 'fish suppers'—'supper' being Scottish & Northern Irish chip-shop slang for a food item accompanied by chips. Suppliers in Devon and Cornwall regularly offer pollock and coley as cheap alternatives to haddock due to their regular availability in a common catch. As a cheap, nutritious, savoury and common alternative to a whole piece of fish, fish-and-chips shops around the UK supply small battered rissoles of compressed cod roe.

In Australia, reef cod and rock cod (a different variety from that used in the United Kingdom), barramundi or flake (a type of shark meat), or snapper are commonly used. From the early 21st century, farmed basa imported from Vietnam and hoki have become common in Australian fish and chip shops. Other types of fish are also used based on regional availability.

In New Zealand, snapper was originally the preferred species for battered fillets in the North Island. As catches for this fish declined, it was replaced by hoki, shark (marketed as lemon fish) and tarakihi. Bluefin gurnard and blue cod predominate in South Island fish and chips.

In the United States, the type of fish used depends on availability in a given region. Some common types are cod, halibut, flounder, tilapia or, in New England, Atlantic cod or haddock. Salmon is growing common on the West Coast, while freshwater catfish is most commonly used in the Southeast.

### *Accompaniments*

In chip shops in the United Kingdom and Ireland, salt and vinegar is traditionally sprinkled over fish and chips at the time it is served.

Suppliers use malt vinegar, onion vinegar (used for pickling onions), or the cheaper non-brewed condiment.

***Figure:*** *Cod and chips in Horseshoe Bay, Canada, served with a lemon wedge and tartar sauce*

In Britain a portion of mushy peas is a popular side dish as are a range of pickles that typically include gherkins, onions and eggs. In table-service restaurants and pubs, the dish is usually served with a slice of lemon for squeezing over the fish and without any sauces or condiments, with salt, vinegar and sauces available at the customer's leisure.

In Ireland, Wales and Northern England, most takeaways serve warm portions of side-sauces such as curry sauce, gravy or mushy peas. The sauces are usually poured over the chips. In some areas, this dish without fish is referred to as 'wet chips'. Other fried products include 'scraps' (also known as 'bits' in Southern England or 'batter' in North-East England), originally a by-product of fish frying. Still popular in Northern England, they were given as treats to the children of customers. Portions prepared and sold today consist of loose blobs of batter, deep fried to a crunchy golden crisp in the cooking-fat. The very popular potato scallop or potato cake consists of slices of potato dipped in fish batter and deep fried until golden brown. These are often accompanied for dipping by the warm sauces listed above.

In Edinburgh and the Lothians salt and sauce (or *saut an sauce*) is the normal accompaniment traditionally sprinkled over fish and chips or almost anything else bought from the fish-and-chips shops.

The watery "sauce" is a mixture of malt vinegar or non-brewed condiment and/or water and Rowat's or Gold Star brand brown sauce, and it is mixed and bottled—often in an old glass fizzy drink bottle with a hole pierced in the screw cap—by each fish-and-chip shop to their own secret recipe.

In Australia and New Zealand, plain salt is usually sprinkled over fish and chips just before serving. Some customers may choose to salt food themselves, given current public health concerns about salt intake. Another popular condiment is tomato sauce. Tartar sauce is also very popular for the fish. Both tomato and tartar sauce are usually sold in small plastic tubs on the shop counter. Complementary slices of lemon are sometimes served with the dish or takeaway pack, although many New Zealanders grow lemons at home. Less commonly, following British and Irish traditions, malt vinegar is the condiment of choice of some Australasian fish and chip lovers. In New Zealand, most dish and chip shops now offer a choice between battered and crumbed fish, and battered hotdogs, battered potato fritters and other deep fried novelties are also usually available.

In Canada, fish and chips may be served with the traditional salt and vinegar, but a lemon wedge and tartar sauce is often the accompaniment found in table service restaurants. Coleslaw of both the vinegared or creamy variety is often interchangeably served as a side.

In the United States, most restaurants serve fish and chips with tartar sauce, ketchup, and coleslaw, although malt vinegar also is sometimes offered, especially at UK-themed pubs.

### Vendors

In the United Kingdom, Ireland, Australia, Canada, New Zealand and South Africa, fish and chips usually sell through independent restaurants and take-aways. Outlets range from small affairs to chain restaurants. Locally-owned seafood restaurants are also popular in many local markets. Mobile "chip vans" serve to cater for temporary occasions. In Canada, the outlets may be referred to as *chip wagons*. In the United Kingdom some shops have amusing names, such as "A Salt and Battery", "The Codfather", "The Frying Scotsman", "Oh My Cod", and "Frying Nemo." In countries such as New Zealand and Australia, fish-and-chip vendors are a popular business and source of income among the Asian community, particularly Chinese migrants.

In Ireland, the majority of traditional vendors are migrants or the descendants of migrants from southern Italy. A trade organisation exists to represent this tradition. Fish and chips is a popular lunch meal eaten by families travelling to seaside resorts for day trips who do not bring their own picnic meals.

Fish-and-chip outlets sell roughly 25% of all the white fish consumed in the United Kingdom, and 10% of all potatoes.

The existence of numerous competitions and awards for "best fish-and-chip shop" testifies to the recognised status of this type of outlet in popular culture.

Fish-and-chip shops traditionally wrapped their product in newspaper, or with an inner layer of white paper (for hygiene) and an outer layer of newspaper or blank newsprint (for insulation and to absorb grease), though the use of newspaper for wrapping has almost ceased on grounds of hygiene. Nowadays establishments usually use food-quality wrapping paper, occasionally printed on the outside to emulate newspaper.

The British National Federation of Fish Friers was founded in 1913. It promotes fish and chips and offers training courses.

A previous world record for the "largest serving of fish and chips" was held by Gadaleto's Seafood Market in New Paltz, NY. This 2004 record was broken by Yorkshire pub *Wensleydale Heifer* in July 2011. An attempt to break this record was made by Doncaster fish and chip shop Scawsby Fisheries in August 2012, which served 33 lb (13.6 kg) of battered cod alongside 64 lb (27.2 kg) of chips.

### Cultural Impact

The long-standing Roman Catholic tradition of not eating meat on Fridays - especially during Lent - and of substituting fish for other types of meat on that day - continues to influence habits even in predominantly Protestant, semi-secular and secular societies. Friday night remains a traditional occasion for eating fish-and-chips; and many cafeterias and similar establishments, while varying their menus on other days of the week, habitually offer fish and chips every Friday.

In Australia and New Zealand, the words "fish and chips" are often used to highlight the difference in each country's short-i vowel sound [j]. Australian English has a higher forward sound [i], close to the *y* in *happy* and *city*, while New Zealand English has a lower backward sound [X], a slightly higher version of the *a* in *about* and *comma*. Hence many people from other dialects hear an Australian say "feesh and cheeps" and a New Zealander say "fush and chups" for fish and chips.

### *Environment*

In the UK, waste fat from fish and chip shops has become a useful source of biodiesel. German biodiesel company Petrotec have outlined plans to produce biodiesel in the UK from waste fat from the British fish-and-chip industry.

### Fish Fry

A fish fry is a meal containing battered or breaded fried fish. It typically also includes french fries, coleslaw, hushpuppies, lemon slices, tartar sauce, hot sauce, malt vinegar and dessert. Some Native American versions are cooked by coating fish with semolina and egg yolk. Fish is often served on Friday nights during Lent as a restaurant special; such a menu offering is sometimes "all you can eat" and occasionally family style (serving dishes brought to and left at the table). Beer is a common beverage of choice to accompany a fish fry. A fish fry may include potato pancakes (with accompanying side dishes of sour cream or applesauce) and sliced caraway rye bread if served in a German restaurant or area.

A Shore Lunch is traditional in the northern United States and Canada. For decades outdoor enthusiasts have been cooking their catch on the shores of their favourite lakes. Fish fries are very common in the Midwestern and northeastern regions of the United States. This is especially true for predominantly Roman Catholic communities on Fridays during Lent, when regulations call for abstinence from most meat products. The modern fish fry tradition is strongest in Wisconsin, where more than 1,000 eateries hold a weekly fish fry on Fridays, and often on Wednesdays. Fish fries there are offered at many non-chain restaurants, taverns that serve food and some chain restaurants.

The Friday night fish fry is a popular year round tradition in Wisconsin among people of all religious backgrounds. A typical Wisconsin fish fry consists of beer batter fried cod, perch, bluegill, walleye, or in areas along the Mississippi River, catfish. The meal usually comes with tartar sauce, french fries or German-style potato pancakes, coleslaw, and rye bread, though baked beans are not uncommon.

The tradition in Wisconsin began because Wisconsin was settled heavily by German Catholics whose religion forbade eating meat on Fridays. The number of lakes in the state meant that eating fish became a popular alternative. Scandinavian settlements in northern and eastern Wisconsin favoured the fish boil, a variant on the fish fry, which involves heating potatoes, white fish, and salt in a large cauldron.

### *Northeastern United States*

Battered or breaded haddock and cod fish fry is one of the trademarks of upstate New York cuisine and northwestern Pennsylvania, especially Buffalo, as well as Rochester, Albany, Syracuse, New York, and Utica, New York. The majority of restaurants in these cities serve a fish fry on Friday, even outside Lent, and it's often available throughout the week.

### *Southeastern United States*

In the southern United States, a fish fry is a family or social gathering, held outdoors or in large halls. At a typical fish fry, quantities of fish (such as bream, catfish, flounder and bass) available locally are battered and deep fried in cooking oil. The batter usually consists of corn meal, milk or buttermilk, and seasonings. In addition to the fish, hushpuppies (deep fried, seasoned corn dumplings), and cole slaw are served. These events are often potluck affairs. In Georgia and South Carolina, fish are dipped in milk, then into a mix of flour, cornmeal and seasonings before frying. Buttered grits is often a side dish.

### *Knights of Columbus*

The Knights of Columbus have long held fish fries on Fridays. They are often advertised as family events, as socials for parishioners.

## Fried Fish

Fried fish refers to any fish or shellfish that has been prepared by frying. Often, the fish is covered in batter, or flour, or herbs and spices before being fried. Fish is fried in many parts of the world, and fried fish is an important food in many cuisines. For many cultures, fried fish is historically derived from *pescado frito*, and the traditional fish and chips dish of England which it inspired.

The latter remains a stable take-out dish of the UK and its former and present colonies. Fried fishcakes made of cod (and other white fish, such as haddock or whiting) are a widely produced seafood available in the frozen food sections of U.S. grocery stores and at fast-food restaurants, such as McDonald's Filet-O-Fish. Long John Silver's, Skipper's, Captain D's, and Arthur Treacher's are well-known North American chain restaurants that serve fried fish as their main food offering. Catfish are also a prevalent farm-raised type of fish that is often served fried throughout the world. A classic fried fish recipe by the French is the *Sole meunière*.

Community fish fries are popular in the southern region of the United States. These social gatherings may centre around a church, a civic organisation or serve as a fundraiser for a club, volunteer fire department, a school or other organisation. In the U.S., especially the Upper Midwest, the Northeast, and the Mid-Atlantic states, community fish fries are slightly popular, sometimes having a religious connection when held in church basements or lots, in observation of Lent, for example.

A fish fry is generally an informal. A "shore lunch" is a tradition in northern U.S. and Canada, where outdoor enthusiasts cook their catch on the shores of the ocean, or lake depending where the fish was caught.

### Fried Fish Dishes

| *Name* | *Description* |
|---|---|
| Pescado frito | An Andalusian dish, made by coating the fish (blue or white fish) in flour and deep fried in olive oil then sprinkled with salt as the only seasoning. Spanish Jews brought the recipe to England during the 17th Century, helping the eventual development of Fish and chips |
| Fish and chips | Battered fish which is deep-fried and served with chips. A popular take-away food in the United Kingdom, Ireland, Australia, New Zealand, Canada and South Africa. |
| Fishcake | A fishcake or fish cake consists of filleted fish and potato patty, sometimes coated in breadcrumbs or batter, and fried. They are similar to a croquettes and are often served in British fish and chip shops. |
| Fish fry | Contains battered or breaded fried fish. It is typically accompanied with french fries, coleslaw, hushpuppies, lemon slices, tartar sauce, malt vinegar and dessert. |
| Fish finger | A processed food made using a whitefish, such as cod, haddock or pollock, which has been battered or breaded. They are known as fish sticks in North America. |
| Fried prawn | Popular in Japan where it also used as an ingredient of bento. |
| Fried shrimp | Batter coated and deep-fried shrimp is usually cooked in vegetable oil |
| Tempura | Japanese dish of seafood or vegetables that have been battered and deep fried |

*Contd...*

| ***Name*** | ***Description*** |
|---|---|
| Whitebait fritter | Whitebait is a collective term for the small fry of fish. These are tender and edible, and can be regarded as a delicacy. The entire fish is eaten including head, fins and gut. Some species make better eating than others, and the particular species marketed as "whitebait" varies in different parts of the world. In New Zealand whitebait fritter are popular. Whitebait is combined with eggs or egg white, and cooked as an omelette is cooked. |

## Pescado Frito

*Pescado frito* (literally, "fried fish"), or *Pescaíto frito* (Andalusian dialect), is a traditional dish from the Southern coasts of Spain, typical in Andalusia, but also found in Catalonia, Valencia, the Canary Islands and the Balearic Islands. It is also consumed as a delicacy in inland Spain, being very common in the inland Andalusian provinces of Seville and Cordoba. It is also very common throughout the Mediterranean Sea, Provence (France), Roussillon (France) and in the coastal regions of Italy (where the most common variant using salt cod fillets is known as *filetto di baccalà*) and Greece. It was also eaten by the Romans in ancient Rome.

It is made by coating the fish (blue or white fish) in flour and deep fried in olive oil then sprinkled with salt as the only seasoning. It is usually served hot, freshly fried, and can be eaten as an appetizer (for example with a beer or wine), or main course. Usually, it is served with fresh lemon, which is squeezed over the fish or occasionally in "escabeche" (a vinegar marinade/souse).

It is also a traditional Shabbat fish dish (usually cod) originating amongst the 16th century Andalusian Jews of Spain and Portugal. The deep-frying of the fish in vegetable oil makes it crisp and light even when eaten cold, and it is a favourite dish of the late breakfast or lunch after synagogue services on Saturday morning.

There is a general belief that pescado frito was possibly an inspiration for the English fish and chips.

# Bibliography


Atre, P. K.: *Fish Genetics and Aquatic Environment*, Navyug, Delhi, 2008.

Balfour, F. M.: *A Monograph on the Development of Elasmobranch Fishes*: Macmillan, London, 1878.

Bardach, J. E., and Villars, T.: *The Chemical Senses of Fishes*, Academic Press, London, 1974.

Barg, U., and M. J. Phillips: *Environment and Sustainability*, Food and Agriculture Organization of the United Nations, Rome, 1990.

Bone, Q.: *Fish Physiology*, Academic Press, New York, 1999.

Bowman, S.J.: *Digest of statistics of IAFMM*, Potters Bar, U.K., IAFMM, 1984.

Breder, C. M., and Rosen, D. E.: *Modes of Reproduction in Fishes, How Fishes Breed*, Natural History Press, New York, 1966.

Bull, H. O.: *The Physiology of Fishes*, Academic Press, New York, 1989.

Chhapgar, B. F.: *Fishes of India*, Oxford University Press, Delhi, 2008.

Chow, Ching Kuang: *Fatty Acids in Foods and Their Health Implications*, Routledge Publishing, New York, 2001.

Deka, Manab: *Fish Fermentation: Traditional to Modern Approaches*, New India Pub, Delhi, 2009.

Dholakia, A.D.: *Fisheries and Aquatic Resources of India*, Daya, Delhi, 2004.

Erasmus, Udo: *Fats That Heal, Fats That Kill*, Alive Books; Burnaby (BC), 1993.

Frank, C. Edminster: *Fish Ponds for the Farm*, Agrobios, Delhi, 2004.

Geran, James: *The Proper Care of Goldfish*, TFH Publications, NJ, 2000.

Hall, G.M.: *Fish Processing Technology*, Springer, Delhi, 2009.

Hasler, A. D.: *Orientation and Fish Migration, in Hoar, W. S., and Randall, D. J.*, New York, Academic Press, 1971.

Jee, Chandrawati and Shagufta: *Fish Biotechnology*, A.P.H. Pub, Delhi, 2010.

Kulkarni, G.K.: *Fisheries and Fish Toxicology*, APH, Delhi, 2006.

Kumar, Arvind: *Fish Biology*, APH, Delhi, 2005.

Malvee, Sangeeta: *Fish Genetics*, SBS Pub, Delhi, 2008.

Marshall, N. B.: *The Life of Fishes*, Weidenfield and Nicholson, London, 1965.

Moyle, PB and Cech, JJ: *Fishes, An Introduction to Ichthyology*, Benjamin Cummings, NY, 2003.

Nash, Colin (2011) *The History of Aquaculture* John Wiley and Sons. ISBN 9780813821634.

Nelson, J. S.: *Fishes of the World*, John Wiley & Sons, Inc.. , NY, 2006.

Petr, Tomi: *Fisheries in Irrigation Systems of Arid Asia*, Daya, Delhi, 2007.

Puri, Neelima: *Fish Endocrinology*, Vista International Pub, Delhi, 2008.

Rebelin W.E.: *The Pathology of Fishes*, The University of Wisconsin Press, UK, 1975.

Robertson, D.R.: *Fishes of the Tropical Eastern Pacific,* University of Hawaii Press, Honolulu. 1994.

Schmida, Gunther E.: *Rainbow Fish*. Barron's Educational Series, NY, 2000.

Silva, De: *Fish Nutrition in Aquaculture*, Springer, Delhi, 2001.

Srivastava, C. B. L.: *Fish Biology*, Narendra, Delhi, 2008.

Stoskopf, M.K.: *Fish Medicine*, W.B. Saunders Co. 1993.

Trevor, A. Anderson: *Fish Nutrition in Aquaculture*, Springer, Delhi, 2009.

Van Holde KE, Mathews CK: *Biochemistry*, Benjamin/Cummings Pub. Co., Menlo Park, Calif, 1996.

Ward, Barbara E.: *Varieties of the Conscious Model: Fishermen of South China*, Tavistock, London, 1965.

Warner, William W.: *Beautiful Swimmers: Watermen, Crabs and the Chesapeake Bay*, Little, Brown, Boston, 1976.

# Index

❑❑❑